Agile Strategy Management

Techniques for Continuous
Alignment and Improvement

ESI International Project Management Series

Series Editor

J. LeRoy Ward, Executive Vice President

ESI International, Arlington, Virginia

Agile Strategic Management: Techniques for Continuous Alignment and Improvement
Soren Lyngso • 978-1-4665-9607-8 • 2014

PgMP® Exam: Practice Test and Study Guide, Fourth Edition
Ginger Levin and J. LeRoy Ward • 978-1-4822-0135-2 • 2013

PgMP® Exam Challenge!
Ginger Levin and J. LeRoy Ward • 978-1-4822-0208-3 • 2013

PMP® Exam: Practice Test and Study Guide, Ninth Edition
Ginger Levin • 978-1-4822-0224-3 • 2013

PMP® Exam Challenge! Sixth Edition
J. LeRoy Ward and Ginger Levin • 978-1-4665-9982-6 • 2013

Determining Project Requirements, Second Edition: Mastering the BABOK® and the CBAP® Exam
Hans Jonasson • 978-1-4398-9651-8 • 2012

Team Planning for Project Managers and Business Analysts
Gail Levitt • 978-1-4398-5543-0 • 2012

Practical Project Management for Building and Construction
Hans Ottosson • 978-1-4398-9655-6 • 2012

Project Management Concepts, Methods, and Techniques
Claude H. Maley • 978-1-4665-0288-8 • 2012

Program Management Complexity: A Competency Model
Ginger Levin and J. LeRoy Ward
978-1-4398-5111-1 • 2011

Project Management for Healthcare
David Shirley • 978-1-4398-1953-1 • 2011

Managing Web Projects
Edward B. Farkas • 978-1-4398-0495-7 • 2009

Project Management Recipes for Success
Guy L. De Furia • 978-1-4200-7824-4 • 2008

Building a Project Work Breakdown Structure: Visualizing Objectives, Deliverables, Activities, and Schedules
Dennis P. Miller • 978-1-4200-6969-3 • 2008

A Standard for Enterprise Project Management
Michael S. Zambruski • 978-1-4200-7245-7 • 2008

The Complete Project Management Office Handbook, Second Edition
Gerard M. Hill • 978-1-4200-4680-9 • 2007

Agile
Strategy
Management

Techniques for Continuous Alignment and Improvement

Soren Lyngso

ESI™
INTERNATIONAL
an **informa** business

CRC Press
Taylor & Francis Group
Boca Raton   London   New York

CRC Press is an imprint of the
Taylor & Francis Group, an **informa** business
AN AUERBACH BOOK

CRC Press
Taylor & Francis Group
6000 Broken Sound Parkway NW, Suite 300
Boca Raton, FL 33487-2742

© 2014 by Taylor & Francis Group, LLC
CRC Press is an imprint of Taylor & Francis Group, an Informa business

No claim to original U.S. Government works

Printed on acid-free paper
Version Date: 20130916

International Standard Book Number-13: 978-1-4665-9607-8 (Hardback)

Library of Congress Cataloging-in-Publication Data

Lyngso, Soren.
   Agile strategy management : techniques for continuous alignment and improvement / Soren Lyngso.
       pages cm. -- (ESI international project management series ; 18)
   Includes index.
   ISBN 978-1-4665-9607-8 (hardcover)
   1. Information technology--Management. 2. Strategic planning. I. Title.

HD30.2.L95 2014
658.4'012--dc23   2013035796

Visit the Taylor & Francis Web site at
http://www.taylorandfrancis.com

and the CRC Press Web site at
http://www.crcpress.com

Contents

List of Figures

Preface

My companies have had the opportunity to facilitate and coach major industrial, public, and financial organizations with their change of direction and improvement of their business, i.e., their strategic initiatives.

The book makes it possible for me to share our experience with you.

Reading the book in connection with strategic initiatives that you get involved with you'll have access to concrete guidelines and techniques that can prevent major blunders and even improve your results in many cases.

The strategic initiatives that we have been involved with were based on business needs to improve efficiency, competitiveness, growth, profitability, market presence, and other strategically important measures of their success.

We implemented the solutions in the form of concrete improvements to production, logistics, client service, IT etc., supported by or implemented as efficient information systems.

The book presents a number of challenging situations, where my companies had the opportunity to facilitate and coach our clients to be successful with their strategic initiatives. In all cases, the book explains why we handled the cases the way we did.

I believe the book can contribute to prevent at least some problems with strategic initiatives in public and private organizations of all sizes.

Our facilitation and coaching were mere catalysts and in most cases, our contribution to the success of our clients has been invisible once the required result was in place.

This is how we want it to be!

The focus of the book is on the other hand our experience from involvement as facilitators and coaches to be shared with you.

I wish to thank former and current employees, clients, and sponsors for their great help in making this book a success. Especially I would like to mention the support from Jesper Ringvad Nielsen (now working for Deloitte in Luxembourg), Claus Raa, Danmarks Nationalbank, Jan Hallberg, Ericsson Sweden, my editor, and John Wyzalek of Taylor & Francis.

Introduction

The book presents some of the cases where we have facilitated and coached major clients to reach important business objectives. The focus is for once on our involvement as facilitators and coaches. Nonetheless, it also mentions the type of results obtained by the clients and the challenges leading to our involvement.

Our principles for agile strategic management and documented standards made the service rendered a sound long-term investment for our clients.

People, organization, and communication are the pivot points of our methods. Building teams with people who can and will accomplish exceptional results is great fun as well as a great challenge that has made our effort worthwhile.

In order to support the motivation of the people involved with strategic initiatives we have used processes of teambuilding and solution implementation that make their contribution so visible that they want to take ownership of intermediate and final results.

I believe the book can contribute to prevent at least some problems with strategic initiatives in public and private organizations of all sizes.

The real life experiences that the book contains are not always success stories. It also tells about major blunders and about how to avoid the pitfalls that one meets while governing strategic initiatives.

It will please me if you find the book inspiring and even fun reading.

It is not the ultimate way to ensure a high quality strategy—just some ideas based on experience from strategic initiatives that my companies or I have managed, coached, and facilitated based on our specific methods.

You can use the book to look up "how to" examples chapter by chapter that explain various ways to use the agile methods and techniques that we have experienced; e.g.:

Chapter 1 presents an overview of what strategic initiatives and agile strategy management is about.

Chapter 2 presents how strategic initiatives can be organized for optimal communication and agile governance.

Chapter 3 presents how the detailed objectives of strategic initiatives can be defined for optimal program and project management.

Chapter 4 presents how to procure competent solution providers in such a way that you can ensure the quality of the solutions they deliver.

Chapter 5 presents how you can implement solutions that ensure the success of your strategic initiatives.

Chapter 6 presents how you can ensure the long-term benefits of your strategic initiatives.

Chapter 7 gives you an overview of the main conclusions and ideas presented in the book.

In all chapters you find pertinent business cases that explain how to establish organizations for change and how to ensure that these intermediate organizations stay motivated until final solution delivery.

The quality management methods that contribute to the agility of the methods and techniques will probably capture the interest of corporate quality and strategy managers, and implementers of solution components.

Other ways you can use the book are:

Managers on all levels can read the book for inspiration—and hopefully be amused.

Some organizations might want to implement the presented methods, techniques, and standards as part of their own methodology.

The book is not a textbook because it does not pretend to tell you how to do your job; but it does show you what we have done under specific circumstances that are probably not very different from what you will meet in your job.

I hope the book will give inspiration to leaders, managers, and other people in organizations who get involved with and who want to maintain or improve their current corporate strategy through strategic initiatives.

It will please me if the ideas and experiences that are presented here with real life case stories are relevant to your needs.

About the Author

Soren Lyngso, MA (economics), Copenhagen University, PMP, general manager, facilitator, and coach, has more than 30 years of experience in general, project, and program management and from implementation of complex business and IT solutions in different industries and countries.

Experience has been gathered from Danish and international projects spanning maintenance of oil production platforms, factory implementation in the pharmaceutical industry, container line system implementation, distribution of toys, cash card system implementation, defense facility management, and in private banking and asset management solution implementation.

Currently his main occupation is teaching and coaching project and program managers on all levels in Europe and the Middle East—the favorite subject being "Rapid Assessment and Recovery of Projects in Trouble." He has published his quality model for strategically aligned solution implementation on his web site www.lyngso.lu.

He lives in Luxembourg with his wife, who is a manager in the French Public Administration.

1

Strategy Quality and Strategy Success

This chapter explains why and how a strategy can be quality managed to deliver successful results.

The book does not wipe out the possibility to pursue a tacit strategy, i.e. a strategy living inside somebody's head and not documented anywhere; on the contrary, it points out that anyone with a little luck can have a lot of success simply by doing what they believe is right. The problem with a strategy based on pure luck is that we cannot learn from it, there is no way we can repeat the process that leads to the high quality strategy.

In order to be able to quality manage a strategy; such a strategy must be established and implemented based on sound and visible processes that can be documented, evaluated, improved, and (re-)used.

The processes described in the book with their organization, standards, and results cover the full strategy lifecycle from establishment to successful implementation or abolition.

1.1 KNOWLEDGE SHARING

The initial knowledge in my companies was based on more than 10 years of practice in project and program management governing information system development and implementation in support of strategic business change and development initiatives.

We have never regarded our methods, techniques, and standards as proprietary or confidential. There are a few reasons for this attitude:

- Once an idea is born, it is the foundation for even better ideas—if you share it.
- We want to share our knowledge with our clients and other partners.

- We do not regard our methods as unique or the best, but we want them to be world class and at least as good as the best.
- We learn as much from clients and partners as they learn from us.
- We are here to have fun, not to live in fear that someone will steal our ideas.

In order to be able to transfer our knowledge, we had to document our methods and techniques in such a way that we could teach it.

All the employees in my companies were encouraged to teach even if they had not tried it before. After all, we were introducing new ways to improve business and manage change and quality in corporations, so without the capability to teach we were not able to transfer our knowledge fast enough to the clients and to new employees.

Based on this need for knowledge transfer and sharing, our methods of agile quality management and our agile principles in general were developed under the name of the Lyngso Model (Figure 1.1), which is a comprehensive collection of methods and techniques for the conduction of strategic initiatives.

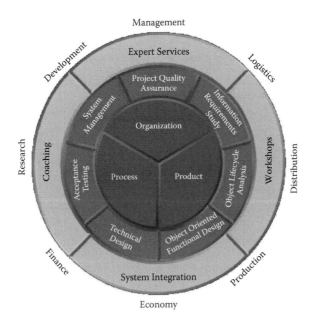

FIGURE 1.1
Lyngso method and technique framework.

All our clients had a copy of the manual that explains how we use the methods, techniques, and standards free of charge. In this way, they could verify what the ideas were behind the work performed by our coaches and facilitators. It was our way to present our quality management system to the clients.

The development of our methods and techniques has never stopped.

Over time, many strategy management standards have been developed, for example, the IS0 9000 public family of standards, the continuous quality improvement with Six Sigma from Motorola, the Capability Maturity Model Integration (CMMI) concerned with organizational method maturity from Carnegie Mellon's Software Engineering Institute (SEI), COBIT, and ITIL.

We continually verify that our agile methods comply with these standards. This is not as difficult as one might think because of the following:

- Our standards are based on specific methods and techniques, which is basically never the case with the public standards that make reference to best practice, but do not show what this best practice is.
- The public standards advise you to establish methods and techniques for the type of work that your organization does, but not how to do this.
- We deliver services based on our specific standards that are continuously improved from working experience.
- We present why and how we have done something. The public standards present what to do and to some extent why, but only in rare cases how.

1.2 SYNCHRONIZATION

Enterprise Information Systems, Business Behavior, and Business Organization are continuously kept synchronized and adapted to the environmental conditions and opportunities in order for the enterprise to obtain the maximum benefits from technology, market opportunities, knowledge, and experience.

If this synchronization is not done, your enterprise will encounter serious problems such as the following:

- Competitive industry will squeeze you out of the market by better usage of technology to offer better products with better performance at the same or lower price.
- Your clients find other more attractive ways to obtain the benefits that you used to offer by replacing your products and services with new ones.

The industry examples are legion. Nokia has lost market share to Apple and Samsung, supermarkets are replaced with shopping centers, European production of basic products such as cloth has been moved to China and India, Novo Nordic has obtained a dominating position on the insulin market based on innovative products, etc.

The needed synchronization is a never-ending process of change. The change is managed in order to ensure that the most competent resources get involved at the right time to produce the solutions that are the best fit to the market conditions and the client needs and expectations, when these solutions are ready for the market.

The synchronization of enterprise Information Systems, Business Behavior, and Business Organization takes place within strategic initiatives (Figure 1.2).

FIGURE 1.2
Synchronized strategy framework.

Strategic initiatives most often comprise planning, development, and implementation of new Information Systems and adaptation to already implemented ones in support of:

- Establishment of new business
- Changes to organization structure
- Implementation of new technology
- Implementation of improved production methods
- Implementation of improved logistics

The information system makes it possible to obtain the required result of the strategic initiative, but the information system is worth nothing if it does not correspond to business processes that deliver real customer benefits.

In the case of the pharmaceutical factory implementation, the first information systems delivered had left quite a few industrial processes under human control, such as, for example, physical and geographical movement of important production elements. When the quality manager had inspected a great number of quality problems, it became clear that 90% of all problems originated in the manual processes. There were no room for such problems in the long run, so today all logistic movements are handled in an integrated process and the error percentage has been reduced to close to zero.

1.2.1 The Importance of Synchronization

Let me tell you about a failed strategic initiative of Information System improvement that ended up in a pure catastrophe because the synchronization of business behavior, organization, and information systems was not done.

A major distributor of big and small electrical household equipment decided to swap its Information System because the current system was getting very expensive to maintain and was very slow.

The IT installation comprised an old-fashioned mainframe with attached PC workstations that the distributor wanted to swap with state-of-the-art Microsoft-based servers with Windows workstations.

The employees and connected clients and partners were used to and very competent in the usage of their current Information System functionality that supported the business processes well nationwide.

The business owner had a friend who was developing and selling a contemporary COTS (Commodity Off The Shelf) system addressing the needs of major equipment distributors. On paper, this new system was based on contemporary IT technology fitting the needs of the major distributor.

As the new COTS application was built for equipment distribution, it was deemed not necessary to establish a requirements specification that would have had to be done by very expensive business consultants.

Therefore, the system was bought from the friend who also became responsible for swapping the business Information System and for training the future users.

The swap consisted of data transfer from the old system to the new one, which was supposed to be much better than the old one—although "better" was defined only as "faster response time and a more user-friendly web-based user interface."

An Accept-Test was established after the data had been transferred to the new system. This Accept-Test was a mere demonstration of the new system for the future users thereof. All questions of a critical nature were wiped away with answers such as "This will be available once in operation and once you have been trained in using the new system." The questions and answers were not documented, and the business owner accepted the new system to go into production immediately.

The start of usage happened as a big bang once all data had been transferred from the old system—of course, only the data relevant for the new system—and the system had proved available to all future users technically speaking.

The future users were:

- Shop owners placing orders and receiving equipment to their local warehouse and in response to client orders.
- Central and distributed personnel managing stock, purchase, logistics, and finance.

Once the users opened their wonderful new system, the problems began:

- The products were there, but only once.
- Stock locations were simple sub-structures to the central warehouse without specific pricing, purchasing, and delivery conditions.

This was completely different from the old system, which had been established as a real Information System in support of their specific business strengths and needs.

The new "solution" was based on business conditions dreamed up by the COTS software vendor.

The COTS software vendor offered very expensive adaptations with delivery lead times that were unacceptable. Furthermore, the COTS vendor would not maintain specific adaptations to the COTS software.

Unbelievably, this is a true 2012 story, and the poor employees and shop owners are still struggling with manual adaptations allowing them to do their business only based on homemade spreadsheets even today in 2013.

1.3 WHY OUR METHODS

In 1986, I established my first company, Lyngso Information Industry, with the objective of delivering strategically aligned Information Systems to our clients.

The background for the "Information Industry" part of the name was that we could promise delivery of fully accepted Information System solutions based on safe estimates emanating from standard dialogues with all pertinent stakeholders in the context of fully standardized processes for business analysis and for object-oriented solution design.

These dialogues take place in scenarios that ensure motivation and best possible contribution from all involved stakeholders to such a degree that these stakeholders want to take ownership of the result—collectively and without conflicts.

It is a basic agile principle of our methodology that we strive to make the client take ownership of whatever is delivered.

We will contribute to the client solution by delivering whatever tools, techniques, and solution components we happen to have the competence to deliver. The final solution inclusive of the knowledge transferred or shared belongs to the client.

Our methods and techniques cater to Strategic Initiative establishment and governance that align Information System development, implementation, support, and governance in any industry with corporate strategy.

The methods and techniques have been used successfully in industries such as:

- Oil and gas production
- Logistics
- District heating production and distribution
- Electricity distribution
- Pharmaceutical production
- Public sector
- Healthcare
- Finance (bank and insurance)

Wherever we have contributed to strategically aligned corporate information system solutions we have left our documented standards used for this work with the clients for them to use it without any restrictions.

It is a real pleasure to come back more than 10 years later to find that the standards are still in use in the organization.

Our first clients declared that it was the first time they saw a true strategic angle to Information Technology and Information System solution development and implementation. Today, more than 25 years later, we are not first anymore, and probably not even unique, but we do have some stories to tell (Figure 1.3).

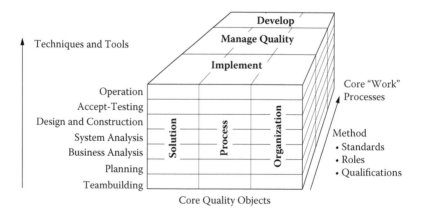

FIGURE 1.3
The strategy management framework.

All method components are explained with respect to the core quality objects:

- Organizational requirements
- Process requirements
- Solution requirements

The Core Work processes of Implementation, Development, and Quality Management are all established in support of agile behavior, where the concurrent involvement of competent resources ensures fast adaptation to unexpected and risk-managed situations and events.

The standard processes and documentation used in the core processes contribute to efficient progress tracking and quality management from the start of a strategic initiative to the delivery of the expected result.

The concrete techniques and tools that are used in all development, implementation, and quality management processes ensure the full traceability of all results from idea to solutions in operation. Traceability is especially important while working agile, where results are adapted to changes in stakeholder demand.

The methods have been developed to be used during the different phases of strategic initiatives, where the strategic initiative can be an information system engineering project; but the methods and techniques have also been used for pure business process engineering such as the establishment of a new factory.

1.3.1 Agile Strategy Quality Management

The techniques and standards for team building and object-oriented business analysis and solution design have been developed to solve the broader and more complex tasks of strategic Information System planning, development, and implementation governed by Agile Strategy Quality Management.

The Agile Quality Management standards used in strategic initiatives comprise:

- Identification and activation of stakeholders to be involved
- Communication
- Team building

- Decision making
- Documentation

The Quality Assurance and the Quality Control methods of Agile Strategy Quality Management ensure that the key-stakeholders are satisfied with the delivered solutions.

The standards are established to be complementary to and sometimes replace components in industry specific or national method standards for delivery of solutions with the quite ambitious objective that:

> The standards can be understood concurrently by the most hard-core technicians, by the top-level visionary leaders, and by all other strategic initiative stakeholders.

Each strategic initiative stakeholder expects different types of benefits from the initiative, and each one reviews and tests the solution to be delivered for his or her own reasons, that is, the proper WHY that you need to understand.

The agile ongoing quality assurance of the solution, the organization, and the processes of a strategic initiative has contributed to the success of the methods used.

A fundamental capability of the methods is that they allow rapid solution development in order to be able to capture the solution benefits while they are relevant. This capability is ensured by early visualization of the complete solution structure and by ensuring that solution elements can be delivered and made productive early during a strategic initiative; that is:

- Complex solutions are broken down into fully functional business solution components without losing the overview of the total solution.
- Early delivery of solution components allows experience to be gathered early for improved estimation, planning, and optimal adaptation to risk events and conditions.
- Early usage of solution components allows the users to gain benefits and to avoid major problems.

In the context of methodology, it should be remembered that no method is perfect and that a great number of different industry, enterprise, or national specific standards exist. While establishing our specific methods we have tried to provide the following advantages compared with such other methods and standards:

- People and communication come before processes, and processes come before documentation standards for the very simple reason that no process can run without resources, and even well-chosen people cannot perform well without good processes to govern their work.
- Documentation standards are simple and cover only the necessary elements. The documentation standards can be seen as content suggestion or as checklists. The methods promoting documentation standards differentiated between complex and simple projects are doomed to fail because they never fit all projects.
- By providing only basic standards and norms, you give the teams the ability to expand to standards of their own that are required for the production of best quality or at least feasible results on their specific tasks. Again, the freedom to act is in focus.
- Your challenge as a leader, manager, or coach is to find the best people and build the best teams and to provide them with standards that work without constraining their ability to perform and adapt the standards to their needs.
- The methods have one basic requirement to all standard documents, which is that they must answer all WHY questions when used, for example:
  - Why has the team been established the way it is?
  - Why is this objective important for the enterprise?
  - Why is this activity conducted the way it is?
  - Why does this solution component function the way it does?
  - Why is this test done?
  - Why is this suggested improvement an improvement?

1.3.2 Quality Management Objects

Quality management comprises three process management classes with explicitly defined organization requirements, procedures, and result standards:

- Quality assurance ensures that a solution will satisfy the stakeholder needs and requirements. This is handled by establishing agreed standards for all resources, procedures, and solution elements to be involved or delivered.

- Quality control is the ongoing effort to maintain the integrity of a method to be able to achieve the required solution quality. We use communication, interview, and review techniques combined with advanced testing and verification procedures supported by standard quality management information systems to perform and document quality control.
- Quality improvement is the purposeful change of a method to improve the reliability of achieving a required solution. We have improved our standards and techniques periodically based on Lessons Learned.

For each quality management process class we use three quality objects as a foundation for evaluation of the performance of this process class:

- The Solution object defines the properties that are required from the solution to be delivered.
- The Process object defines the properties that are required for high performance delivery of the required solution.
- The Organization object defines the properties that are required for efficient communication, competence establishment, and decision making.

Our work with clients and partners has allowed a continuous improvement of our standards, techniques, and tools based on a vast base of gathered and documented experience and knowledge, some of which I want to share with you.

1.4 STRATEGY AND STRATEGIC INITIATIVES

1.4.1 The Strategy

Corporate leadership establishes the strategy of a corporation. The strategy tells you why the organization has been established the way it is:

- Organization structure and geographical locations
- Products
- Business operations

The strategy comprises:

- The corporate vision statement that "paints" a picture of how the corporation would like to be observed and how it observes itself in the future.
- The corporate mission that tells a story about how the corporation intends to contribute to the happiness of its stakeholders. This is the strategy quality objective.
- The confidential corporate objectives known by and sometimes contractually committed to by management tells you the direction followed by the corporation, for example:
  - Internationalization
  - Growth by acquisition
  - Profitability (Return on Investment, Return on Equity)
  - Sustainability
  - Technological superiority

All organizations whether public or private have a strategy and perform business activity governed by this strategy more or less successfully.

Key performance indicators (KPI) and benchmarks measure the strategy quality and success.

Corporate management translates the strategy into detailed organizational constructions, business procedures, and strategic initiatives that can ensure and improve the strategy quality.

1.4.2 The Strategic Initiatives

The strategic initiatives establish the WHY, the WHAT, the WHEN, the HOW, and the WHO concerned with sustaining, changing, and improving business procedures and infrastructure in support of the corporate strategy.

Strategic Initiatives usually comprise information systems establishment or improvement in support of business operations. When my companies have been involved with corporate strategic initiatives, this has always been the case.

Strategic Initiatives are programs or projects.

1.5 AGILE PRINCIPLES

The agile principles of our methods framework ensure that the strategy stakeholders are happy all the time.

Unfortunately, happiness is not the nature of Strategic Initiative stakeholders and especially not when corporate strategy and Information System implementation is concerned. Most often conflicts and disagreements between owners, operators, and unions about how to handle and manage the strategic initiative pave the way for mistrust and insecurity.

There is a long history of failed strategies, missed opportunities, and Information Systems that never were delivered, disasters that unfortunately enough support the normal stakeholder suspicion and mistrust.

On top of these bad experiences, the normal resistance to change and the classical differences in objectives between management and employees pave the way for unhappy stakeholders.

The agility is there to overcome the constraints of mistrust and suspicion among strategic initiative stakeholders. The core agility principle is:

> Each process from the definition of the initial need for change to the final delivery of the agreed solution contributes to stakeholder trust, mutual respect, motivation, and willingness to take ownership of the solution components delivered.

This is the philosophy behind our processes, tools, and techniques. They involve the stakeholders in such a way that they feel that the results obtained belong to them. In this way, each process contributes to the motivation for the next one.

People creating software established the initial agile principles ("The Agile Manifest") that you can look up on the web. This limits its formulation to very specific software development activity, which will not cover the needs of strategic initiatives and Information System development and implementation that involve many other elements than software and software development.

Our principles of Agile Strategy Management comprise the following:

> The highest priority is to satisfy the stakeholders through early and continuous delivery of valuable solution components. We understand why the stakeholders need the solution and we know that the stakeholders like the solution.

Working solution components are the primary measures of progress.

Working solution components are delivered frequently, from a couple of weeks to a couple of months, with a preference to the shorter timescale. The delivery process is Simulated Accept-Testing where the developers of the solution present it to future users, who discuss it with the developers and test it out after thorough education and support.

Changing requirements are welcome, even late in development. The agile processes harness change to obtain improved stakeholder benefits.

Development, Implementation, and Quality Management stakeholders work closely together throughout the project. This is ensured by the environment of technology, communication facilities, space, and more. One of our basic requirements is a War Room where the involved parties assemble all that is needed to generate and evaluate the solutions required for decision making.

Projects are built around motivated individuals. It is ensured that they have the environment and support they need. As the individuals in the project teams have been selected based on their skill, experience, and competence, we can and will trust the teams to get their jobs done. This is the "no excuse for failure" principle. If the team claims that it needs something to get the job done, then it gets that something.

We do not establish or accept conflict during solution implementation; instead, we ask for a mutually agreed solution and progress without further discussion. Once the solution is in place, we are happy to evaluate if something could have been done differently and better.

The most efficient and effective method of conveying information to and within a development team is regular face-to-face conversation. One or more War Rooms cater to this behavior. We have even gone so far as to establish a contract-based situation where software development and solution implementation work was allowed to take place only in the War Room. No solution component could be brought in from the outside and nothing produced inside could leave the War Room before it had been fully Accept-Tested and signed off for IT and business operation. The War Room was a set of containers with approximately 500 m^2 of spaces for work, education, and testing fully and securely supported by the central IT infrastructure.

Agile processes promote sustainable development. This principle ensures that processes can be repeated and evaluated for possible improvement.

Continuous attention to technical excellence and good design enhances agility. Time is devoted to ensure that state-of-the-art technology is used, which again implies that the technology used is a solid foundation for the solutions that use it. Good design ensures normalized processes and data, which ensure integrity and valid integration in support of the agreed business needs. Once the business needs change it is easy to adapt good design, while bad design reveals itself under such conditions.

Only agreed necessary work is done. We list work not to be done explicitly only if doubts are raised. This is the Lean principle to avoid doing not required or not needed work.

The best architectures, requirements, and designs emerge from self-organizing teams. The self-organizing does not ensure this situation, but the fact that teams have been put together in such a way that self-organizing is possible. We ensure that teams can:

- Make decisions
- Initiate work
- Do the work
- Evaluate the progress and the results

This makes simulated Accept-Testing great fun because the delivered solutions do work, although they may be improved in a meaningful (Lean) way.

At regular intervals, the teams evaluate how to become more effective, then tune and adjust their behavior accordingly. The evaluation comprises communication between teams and their environment of resource providers and other key-stakeholders.

The teams have fun:

- We celebrate visibly our successes, also the small ones.
- We do not hesitate to show appreciation of others.
- We are not jealous, but we like a good fight.
- An individual achievement is a team victory.

The fun part can comprise games that require professional knowledge and, quite often, professional development. The games provoke friendly competition in a team and between teams. The games are a great way to get to understand and respect the value of different personalities in a team. The teams do not need rules of the game imposed from outside the team, but they can easily adapt to such rules if needed.

1.6 SCOPE ESTABLISHMENT

A strategic initiative can be concerned with reaching many different objectives regarding different business situations.

This initial set of objectives provided by corporate leaders is an indication of what kind of business and stakeholders that must be involved in the establishment of the strategic initiative. In most cases in which we have been involved, these objectives were not precisely defined to establish the full scope of the strategic initiative on their own.

Only when we have asked key-stakeholders about their points of view on the objectives do we get a more precise idea about the scope that can comprise elements such as:

- Budget
- Competition
- Environment
- Technology
- Legal issues

Therefore, the first action in the establishment of a strategic initiative scope is to identify and to have open dialogues with potential key-stakeholders to be involved with the initiative, while you respect that you do not know what the real scope is—yet.

In order to identify all stakeholders to get involved or to be communicated with you need to look at the complete value chain (Figure 1.4).

A complete quality assured scope and stakeholder definition will follow later through a well-defined quality managed standard procedure such as Process Quality Assurance (PQA).

The result of the dialogues with the potential key-stakeholders is documented in an invitation to the group of identified key-stakeholders to participate in a PQA workshop.

In the workshop, the participating key-stakeholders define a more detailed scope of the strategic initiative:

- They present their individual visions of the initiative result.
- They present what they think the initiative mission is and how they can contribute to this mission.

FIGURE 1.4
Value chain example (from http://bettyfeng.us).

- They define the initiative success factors and the agreed critical success factors.
- They define the set of activities required to fulfill the success factors.

The PQA workshop establishes mutual respect among the participating stakeholders. The success factors replace the initial objective.

The full scope of the strategic initiative has only been defined when the complete plan for execution and delivery of detailed results has been signed off.

Imagine a toy manufacturer who wants to have a bigger share of the sale per outlet and at the same time have a higher contribution per outlet. The outlets are toyshops, supermarkets, catalogs, web-shops, etc. that make autonomous decisions about how to display and promote the toy manufacturer's products.

What is the scope?

Right, there is no simple answer to this question!

Within this larger scope, you discover a broad range of benefits that can be obtained based on one or more appropriate strategic initiatives.

You also find threats emanating from, for example, differentiating prices between rich and poor countries or from misplacing stock locations giving too high transportation costs.

1.7 STAKEHOLDER IDENTIFICATION

In order to establish a strategically aligned solution, you deal with a multitude of stakeholders representing all the roles directly involved with development, implementation, quality management, usage, governance, etc. of the solution, as well as the not always visible stakeholders that potentially benefit or suffer from the strategic initiative and its solutions.

When establishing a strategic initiative you make a serious effort to get to know all the stakeholders that are concerned, that is, to have a dialogue with key persons and organizations that potentially could benefit or suffer from it. This is especially true for the less visible and less obvious stakeholders such as unions, politicians, government, legal bodies, and potential competitive businesses and partners.

In several cases, leaving out potential key-stakeholders has led to the complete failure of the strategic effort. Some real and recent examples of less efficient stakeholder management are addressed next.

1.7.1 The Balder Case

Scandinavian mythology addresses the stakeholder identification problem explicitly with the Balder case, which most children learn about in Scandinavian schools and very well explains why stakeholder knowledge is crucial to strategy success.

The Vikings told the Nordic myths. Their stories are still told; now in books, poems, and films or as bedside stories. They represent some of the first documented storytelling and explains generation after generation the virtues and the dangers of the Aesir gods, how the world was established, and how the different natural events such as the sun and the moon were generated.

The two primary gods were Odin and Thor. Odin was the wise leader, the strategic thinker, while Thor was the strong manager who used his force to ensure the success of his sometimes fancy ideas.

Balder was the son of Odin and Frigg. He was the most handsome of the Aesir gods and on top of this very intelligent when it came to writing and reading runes. However, he did not like to fight. All gods and humans loved Balder. He was married to Nanna.

One day Balder told Nanna that he had dreamed of his own death. Nanna was very worried and went to Odin, who could see into the future, but Odin could not see anything. Odin got worried and saddled his eight-legged horse Sleipnir to ride to see Voelva, who was buried at the entrance to Hades (the world of death). She might know something. Odin saw a table prepared as if they were waiting for guests in Hades. Odin dug up Voelva and asked her whom they were expecting. She did not want to answer, but Odin pressed her to admit that they were expecting to receive Balder. "He will be killed by an arrow from his brother Hoder's bow," she said.

Odin came back home with the terrible news. Nanna and Frigg were very sad; but Frigg did not give up easily. To prevent that Balder died she went out into the world to ask all things to promise that they would not hurt Balder. She succeeded in doing this.

The Aesir gods were happy that Balder now was safe. They arranged for a big party where Balder would be used as shooting target. The party progressed well and stones, arrows, and even Thor's hammer did not do any harm to Balder.

Loke was a terrible troublemaker god. He wanted to see if he could make more harm to Balder than the others. He disguised himself as an old woman and went to Frigg to ask her if there really were not anything that she had not asked. Frigg admitted that the small mistletoe had not been asked.

Loke cut an arrow out of the mistletoe and gave it to Hoder so that he could shoot with that arrow. Hoder refused because he was blind, but Loke offered to give him a hand. Loke placed the arrow on Hoder's bow and helped him to direct the shot. The arrow went straight through Balder, who died on the spot.

Normally the place for dead gods was Valhalla, a paradise with eternal eating and drinking and friendly fighting; but only if the god had died fighting. As this was not the case with Balder he was doomed to go to Hel in Hades.

The god Hermod road to Hel in Hades, where he met Balder and his wife Nanna, who had died of grief on the funeral pyre of Balder. Hermod asked Hel if there was a chance to get Balder and Nanna back. Hel told him that if all living and dead creatures would cry over Balder they could get him back; but if only one creature refused she would keep them.

The Aesir gods were happy to hear this and arranged for everybody to cry. They succeeded until they passed by a cave with an old woman called Tok. She would not cry. The Aesir gods thought that this attitude was strange and went back to the cave, but Tok was gone. Odin analyzed the situation and declared that it must have been Loke disguised as an old woman. "Go and find him," said Odin.

Loke had built a house on a high mountaintop, where he could see anyone who would attack him, but he did not believe that this would save him in the end. He therefore turned himself into a fish. This was counting without Thor, who finally caught Loke.

Loke was bound to a stone. A poisonous snake was hanged above him to let its poison drip on his head. Loke's wife Sigyn collected the poison in a cup. Every time she emptied the cup, the poison hit the head of Loke and his body shook so hard of pain that the earth was quaking.

Balder and Nanna are still living with Hel in Hades.

How often have we left out a stakeholder from attention or involvement in a process because we thought that this potential stakeholder had no real importance to our project?

1.7.2 The Private Bank Case

I was involved with a large program to automate a private bank giving all clients access to full web banking. Focus was on technology and functionality and all of this was successfully implemented and even Accept-Tested and approved before the disaster was discovered.

The bank clients only got involved to enter transactions after the fully integrated solution was technically functional and had been approved from "Friends and Family Testing."

System usage was expected to reach 3000 transactions per day 3 months after the release date.

The solution never had more than 300 transactions performed by the users in one day and in 90% of the cases, known Friends and Family testers performed these transactions.

Millions of dollars were wasted to such an extent that the stock price fell considerably.

The future users were recognized as stakeholders, but an appropriate dialogue was not established with these stakeholders before it was too late.

1.7.3 The DANCOIN Cash Card Case

Another project my organization was involved with was the development and implementation of a Cash Card in Denmark—the DANCOIN case. Several worldwide-recognized patents came out of this exciting project.

The technical development was a great success. Partners such as banks, credit card facilities, and local transportation were directly involved with development and implementation.

A whole city was set up for Accept-Testing end-to-end of the integrated solution with service providers and central bank cash and transaction cost clearing. This acceptance test was a great success also seen from a publication and advertising point of view with press and television coverage.

So, why did the Danish cash card not have success contrary to basically the same card implemented in Rotterdam, Holland?

The failure to succeed occurred because of lag of communication with the corporate management people from the core stakeholder, the local transportation organization, HT:

- In parallel with the DANCOIN development HT developed its proper card and card reader device for its buses and train stations without coordinating this effort with the DANCOIN project.
- On the eve of going live, HT refused to implement a DANCOIN reader.
- Only inferior usage such as a few parking terminals, a few laundries, and some unmanned newspaper kiosks went into production.

This was not enough to pay off the DANCOIN investment and HT had enough resources to simply write off its part of the investment.

1.8 AGILE TEAM BUILDING

Once we have defined and agreed to the larger scope and the required activities and their leaders and managers, we are ready to build the teams for solution establishment, risk, and change management.

The method to get the teams built and to establish agreement about what the solution will be is the continued PQA process. The initial PQA Workshop invitation and the Workshop itself is the very important initiation of the PQA Process.

PQA is a core element of Strategy Quality Management. PQA ensures a precise scope definition broken down into activities, deliverables, resource requirements, process ownership, and management responsibilities.

The PQA process will most often comprise several cascading PQA Workshops and processes with specific invitations and stakeholders for each one, that is, the activities defined on higher PQA process level are candidates for their own more detailed PQA processes with focus on a partial solution delivery.

PQA is Risk Management based, but the focus is Opportunity rather than Threat.

The PQA point of departure is the client's needs of change in his or her current situation of strength and weakness facing the stakeholders that among many others comprise clients and current and potential competition and partners.

PQA not only documents what the scope is, it also documents why the scope is defined the way it is. This ensures the agility of the PQA defined scope because if any why-case changes, then the scope must change. The organization to change the scope is explicitly defined in the PQA result.

The PQA documentation is dynamic in nature. Change management and periodic evaluation of the scope ensures that the definition of the scope of work that governs the development and implementation of the strategic initiative solutions is up to date at any point in time. For each deliverable solution component, the agreed timing, cost, and organization are precisely defined, and risk is managed.

All deliverables (solution components) are realized through three basic processes:

- Implementation that defines business functional requirements and establishes the foundation for Accept-Testing.
- Development that defines technical functional requirements and technology development, implementation, and usage.
- Quality Management that ensures that business implementers and solution developers work closely together (often face-to-face) and evaluate deliverables on a regular basis that allows for adaptation to changed conditions and gained experience in order to ensure full stakeholder satisfaction with the final solution.

1.8.1 No Excuse for Failure Principle

Much too often we have seen projects and major programs moving along with weakly defined organization of responsibility and activity defined on a level where management is impossible.

Such situations lead to crucial lack of commitment from all involved stakeholders because:

- Results do not show up.
- Results show up too late to be useful.
- Results show up without the quality that was never agreed on or documented.

The "no excuse for failure" principle implies the following:

- Involved human and technological resources are fully qualified to deliver the required and documented solution.
- The involved resources are allocated and committed in such a way that their work is done and that the results are delivered without costly interruption, delay, and cost overrun.

The "no excuse for failure" principle ensures that all involved stakeholders are visibly committed and motivated from start to close out of the strategic initiative.

1.8.2 Simulated Accept-Testing

Communication of real measurable results is performed on a regular basis during Simulated Accept-Testing (SAT), which allows the not directly involved business and development stakeholders to evaluate intermediate results.

SAT performance allows timely and pertinent decisions about required changes to take place without disturbing the progress of complete solution delivery.

The SAT communication ensures that final Accept-Testing becomes a mere formality because all involved stakeholders already know exactly what they can expect from the delivered solution components during final Accept-Testing.

1.9 ASSESSMENT AND RECOVERY OF PROJECTS IN TROUBLE

Whenever my organization is called upon by major international organizations to deliver our core services of coaching and facilitation, their strategic initiatives are in trouble. Most often, they have tried to implement a solution for months just to discover that no progress (except for spending time and money) has been achieved.

1.9.1 Medical Factory Implementation

We got involved in a major program to implement a big factory to produce medical equipment using cheap raw material to deliver an end product of high quality to be used worldwide at a price to be acceptable even to poor people.

The program was run like a project with one project manager facing several internal key-stakeholders with considerable internal power and many external stakeholders with legal and political power. The external stakeholders were delivering:

- Buildings
- Production machinery
- QA equipment
- Logistics equipment
- Internal machine control information systems
- Administrative information systems based on the SAP Enterprise Resource Planning COTS software package

The internal key-stakeholders were:

- The future factory manager
- The CEO

The work to be done was defined at a very low level (work packages by contractor) and a lot of work overlapped between contractors.

Arbitration between contractors and project manager was handled at weekly project meetings.

More and more conflicts between all parties surfaced very early. An important reason was that more than one contractor made key decisions and that these decisions were contradictory or at best not visibly aligned with the overall project objectives. This was caused by weakly defined objectives.

The project was finally declared in trouble because major deliveries were slipping without clear responsibility for the delay.

It was obvious that a program organization was needed to govern all stakeholders. Furthermore, a clearly defined unambiguous requirements specification for each contractor was needed and had to be agreed on by all parties.

Our contributions to this program that has since delivered one of the most successful solutions in the history of the pharmaceutical industry were:

- Establishment of a program organization based on visible high-level objectives (critical success factors) and clearly defined high-level activities that each one was a major project on its own. The organization, the objectives, and the activities were approved by top corporate management that became visibly involved in the program management. The project manager of the building implementation said: "This process should have been used from the beginning to avoid the troubles that started more than a year ago ..."

- Coaching of the contractor responsible for delivery of the complete factory control system (a network of control computers connected to the numerical control on all production and quality control equipment).
- Our coaching comprised the establishment of the detailed project plan. We also delivered the method and coaching to develop the normalized data structure and content, and the normalized process structure common to all control computers
- The normalized data and process structure allowed fast corrections of failures and resolution of problems, and ensured an efficient interface with the SAP-based order and production planning and control environment.
- The normalized data and process structure was used and verified early in Simulated Accept-Testing to prove the efficiency of the solution to be developed and delivered.
- We coached the setup and execution of Simulated Accept-Testing supervised by corporate management (the program management team) that signed off on the solution development and implementation.

Our methods used were:

- Process quality assurance to establish the objectives and a complete project plan that allowed reliable estimation, forecasting, and tracking.
- An Information Requirements Study to identify all core objects with their purpose and usage and to ensure that they were complete with respect to the overall success factors for the program and the detailed success factors for the control system production and implementation.
- The Object Lifecycle Analysis to detail, define, and normalize all data and process objects in such a way that all control systems could be developed and implemented where needed, ensuring full integration among control computers and with external systems (numerical control and SAP).
- Simulated Accept-Testing to prove the feasibility of the data and process structure.

Both the pharmaceutical enterprise and the control system contractor coached by us implemented major method and organization improvements in order to benefit from the methods used also in the future.

1.9.2 Complete Swap of All IT Systems in a Private Bank

The new IT manager in a private bank brought us in to assess the situation (the project quality) of the biggest IT and business development program (only defined as a project) ever implemented in this (private and asset management) bank.

The task was that the old computers with COBOL-based Information Systems had to be swapped out and replaced with new technology. The old COBOL-based Information Systems were interfaced with multiple standalone solutions internally and externally. As most of the old in-house developed core-banking IT solutions were getting more and more inefficient, they were not candidates for swapping.

The corporate management had opted for a COTS standard system implemented on state-of-the-art IBM technology (including DB2 relational database) to be interfaced with the set of standalone solutions (mostly in house development) that they expected to survive and to continue to be used in business after the swap.

This COTS system was bought because a neighbor private bank used it and was relatively happy with its solution.

No requirements specifications were established.

A large number of external consultants and internal employees, mostly from IT, had spent more than 18 months with evaluation of the quality of the old solution components while trying to produce a GAP analysis (as is and to be definitions of the software that had never been documented before!).

There were no usable results from this GAP analysis.

Very expensive COTS vendor consultants were working on setting up the COTS system in the private bank even before complete requirements to solution infrastructure, security, safety, and technical environment had been defined. This was, of course, a complete waste of time and money.

In parallel with the GAP analysis, the future end users (bank employees and a few IT employees) were "trained" in the new COTS software that had not been adapted to their requirements (these requirements did not exist).

The users from business and IT were all deeply unhappy and frustrated with what they saw during training and demonstrations. This sporadic and very expensive training provided by the COTS vendor created more frustration and mistrust than learning.

IT management declared the project in deep trouble, stopped all GAP analysis and training activity, and laid off all involved external consultants. During this initial clean-up process, the relatively innocent internal project manager was removed from the project.

Once on board, we established a strategy to recover the project.

This meant, among other major changes, to involve the business organization deeply in planning and requirements specification, while preparing it for future solution implementation, evaluation, and testing.

In order to avoid the risk of losing a lot of capital on activity that does not deliver visibly useful results, we re-planned the program and the projects with a focus on real tangible deliverables—it was made Lean.

The deliverables were real solution components that could be developed, interfaced, and thoroughly tested to be satisfactory to the bank and the employees.

We succeeded in procuring external experts from major service organizations that agreed to deliver the solution components on fixed time and cost based on the solution requirements to be produced by the private bank.

The bank management committed to contribute to the solution requirements and the knowledge and the capacity of resources necessary for solution design, development, Accept-Testing, implementation, and operation.

Internal and external responsibilities were clearly defined, but it was also clearly defined that problems were resolved by mutual proactive solution contribution from all parties.

Payment to external sub-contractors could only be obtained after fully accepted delivery of solution components.

The involvement and re-motivation of the very frustrated user organizations inclusive of IT was established through usage of Project Quality Assurance that quickly visualized the key success factors and all the activities required to achieve the success factors. Success factors in this context are solution capabilities, business and work conditions, organizational competence, and events.

The user organizations agreed to produce requirements specifications quickly that could be used as a foundation for procuring solution providers and COTS-based solution components at a fixed price.

We supported the user management with an adapted and combined Information Requirements Study and Object Lifecycle Analysis that resulted in clearly defined use cases (or sprints) that were used as a basis for structuring the complete business requirements exposed to the potential service providers and sub-contractors.

In order to structure, plan, and track the progress of requirements spec production, we defined a very specific method for how to document all pertinent business processes to be supported by the COTS application and the interfaced standalone systems. To this end, we recommended use of a relatively easy workflow documentation standard that complemented our own Information Requirements Study and Object Lifecycle methods. This was relatively well accepted by the involved business users of the future solution because they had accepted that no solution could be delivered without documented requirements.

It was obvious that external experts from the COTS solution vendor and from organizations with broad and deep experience from implementing the COTS solution were required—and fast.

In parallel with the production of the requirements spec, we prepared the solicitation and tendering among potential sub-contractors. No current business process was left out or adapted in the requirements documentation, which minimized the negative impact from risk-exposed changes to known business processes. Suggested improvements to current business processes were allowed to be documented, but they could be used only for inspiration to the developers and implementers, not as requirements. In this way, we were able to present a complete solution process overview and the first complete work flow documentation very fast to the potential internal and external expert organizations that could perform development and solution implementation.

After 3 months of intensive negotiations and contracting, we had a complete project organization in place for development and implementation, where development consisted mainly in setting up and interfacing COTS applications and development of integration components. Implementation consisted of usage documentation preparation, training material preparation, user training, Accept-Testing, and operation setup.

Nine months later the first solution component went into production. Our contribution to this program comprised:

- Coaching and program management based on project management tools such as Project Quality Assurance, Project Planning with Professional Procurement, and Program Management with many stakeholders and many Work Package project managers.

- Coaching of business management with the establishment of the required physical environment for development and implementation with buildings (containers), rooms, networking, security, and IT. This resulted in a huge war room being made available to all program resources with ID card for entry and exit. For confidentiality reasons, involved development resources were only allowed to work on-site in the war room while fully supervised, while implementation resources from IT and business could work in their own environment when they prepared training material and system operation procedures.
- Establishment of the program organization and the project teams with more than 100 resource persons of whom more than 50% were working full time on the program.
- Progress tracking that was performed once a week with all involved project managers.
- Handling of any cross-organizational issues on a day-to-day basis at 8:30 meetings in the war room.
- Functional implementation issues were analyzed using Object Lifecycle Analysis to ensure a complete and consistent solution implementation.
- Simulated Accept-Testing was established between internal future support, external solution implementers, future end users, and, to some degree, internal management.
- The end users were only formally trained once the future solution components were ready for final Accept-Testing and production.

Frustration was avoided, confidence was re-established, and to some extent, the stakeholders were happy. Unfortunately, it was not possible to cope with all technical issues concerned with the bought COTS software, but that is another story.

1.10 STRATEGY QUALITY AND SUCCESSFUL STRATEGY

Quite often, the corporate strategy is confounded with the corporate objectives or the corporate vision, but the objectives and vision cannot be the strategy on their own because the strategy is also the way, the method chosen to meet the objectives and make the vision come true, that is, the strategic initiatives.

Strategy is not confined to decision making on a board or government level, it is just as much based on decisions made by a department, a ministry, a business unit, or a cross-organizational project or program organization performing strategic initiatives.

The way from corporate visions and objectives to governance of the corporate strategy that fulfils the corporate mission and makes the stakeholders happy is outlined in the simple model shown in Figure 1.5—starting from the top and moving down:

High quality strategies have all objects in the model defined, visibly agreed to by the stakeholders, and documented in order to ensure efficient communication of objectives and measures to be taken by all strategy stakeholders.

1.10.1 Strategy Quality and People

People implement the strategy by explicit definition of, and agreement to, all the quality objects shown from initial visions over performance of projects, programs, and business activity to strategy governance.

Vision Objects	Pictures of the future situation, stories about "life" in the new situation, qualities and attributes of future products or other solution component, ….
Mission Objects	Improve health of people worldwide, combat poverty worldwide, promote democracy worldwide, deliver safe transportation of people and goods, ….
Objective Objects	Market share, growth, return on equity, return on investment, shareholder satisfaction, employee motivation, customer satisfaction, product quality, ….
Strategy Quality Objects	Solution Process Organization
Need Objects	Business/change scope Success factors Critical success factors
Action Objects	Programs Projects Business Activities
Strategy Implementation Objects	Organization with business/change leadership and management portfolios of programs, projects, and business activities benefit achievement measurement
Strategy Governance Objects	Key performance indicators Change control board Corporate communication management

FIGURE 1.5

Quality objects from vision to strategy governance.

The choice of the people involved in a strategy is critical for the quality of the results that can be obtained by following the strategy.

Although this seems obvious, the attention to the choice of people to be involved with a strategy and strategic initiatives is much too often very low and based on the first available opportunities among friends and colleagues.

Once the choice of people to be involved in a strategy has been made and accepted, the leaders and managers must wait for major blunders or failures from a bad choice of people before a change for the better can be made.

It is very difficult to replace people on teams, especially the people leading or managing the teams, because the people that are asked to leave quite often interpret this as a personal defeat. The people that stay often regard the replacement as a management failure or weakness, which raises their suspicions of further blunders to come.

1.10.2 Strategy Quality and Risk

The result of strategic decisions is not always the fulfillment of the objectives that originally led to the decisions. When a result is accepted to be better than the objectives originally established, everything is fine—we can talk about a successful strategy, but it is not sure that we are facing a high quality strategy.

> Pure luck is playing an important role in strategy and strategic initiatives because results of strategic decisions are aleatory.

Only when we visibly apply risk management and visibly control the direct outcome of our strategic initiatives from initiation to final implementation and governance can we talk about high quality of the strategy.

You can recognize a high quality strategy by the fact that the results correspond to the objective that has visibly (i.e., documented) been adjusted to what the stakeholders need and expect in the end.

1.10.3 Strategy Quality and Leadership

I will not evaluate leadership on any scale because all leaders—as it also holds true for people in general—have their own style.

There are, however, certain common characteristics of successful leaders, such as their ability to communicate with the purpose to:

- Listen
- Sell an idea
- Motivate others
- Navigate

These communication capabilities allow leaders to drive strategies through to success, even though the final result is quite different from the one envisioned when the original objectives were set.

The corporate leaders envision the initial strategy objectives, but the chosen teams involved with the strategic initiatives define the final objectives. The corporate leaders play an important role as listeners and promoters of change for the better.

One example of an excellent leader is the Danish Prime Minister Jens Otto Krag, who succeeded in getting Denmark into the European Union in 1972 with a very small majority.

Jens Otto Krag said something in 1966 that places him as one of the first documented agile leader personalities with strong navigation capability seen from a modern standpoint:

"You have a point of view, until you take on a new one."

In 1966, many interpreted this expression as outrageous, while others were confirmed in their perception of Jens Otto Krag as a pragmatic politician who knew how to navigate.

The leaders do not make decisions too often; they make sure that the teams involved with their strategic initiatives are established in such a way that they can operate efficiently and make appropriate decisions by themselves within the scope of the initiative.

We will look deeper into team building in the next chapter.

1.11 LESSONS LEARNED

Once an idea is born, it is the foundation for even better ideas—if you share it.

We learn as much from clients and partners as they learn from us.

We are here to have fun, not to live in fear that someone will steal our ideas.

The presented standards are specific methods and techniques used by my companies, which is rarely the case with the public standards that refer to best practice, but do not show what this best practice is.

We present why and how we have done something. The public standards present what to do and to some extent why, but only in rare cases how.

Enterprise Information Systems, Business Behavior, and Business Organization are continuously kept synchronized and adapted to the environmental conditions and opportunities in order for the enterprise to obtain the maximum benefits from technology, market opportunities, knowledge, and experience.

Strategic initiatives comprise most often planning, development, and implementation of new Information Systems and adaptation to already implemented ones.

Corporate leadership establishes the strategy of a corporation. The strategy tells you why the organization has been established the way it is.

The Strategic Initiatives establish the why, the what, the when, the how, and the who concerned with sustaining, changing, and improving business procedures and infrastructure in support of the corporate strategy.

The Agile Quality Management standards used in strategic initiatives comprise:

- Identification and activation of stakeholders to be involved
- Communication
- Team building
- Decision making
- Documentation

The agile ongoing quality assurance of the solution, the organization, and the processes of a strategic initiative has contributed to the success of the methods used.

The agility is there to overcome the constraints of mistrust and suspicion among strategic initiative stakeholders. The core agility principle is:

Each process from the definition of the initial need for change to the final delivery of the agreed solution contributes to stakeholder trust, mutual respect, motivation, and willingness to take ownership of the solution components delivered.

The first action in the establishment of a strategic initiative scope is to identify the initial objective of the sponsor and the initial key-stakeholders to be involved with the initiative. You respect that you do not know what the real scope is—yet.

When establishing a strategic initiative, you make a serious effort to get to know all the stakeholders that are concerned, that is, to have a dialogue with key persons and organizations that potentially could benefit or suffer from it. This is especially true for the less visible and less obvious stakeholders such as unions, politicians, government, legal bodies, and potential competitive businesses and partners.

Once we have defined and agreed to the larger scope and the required activities and their leaders and managers, we are ready to build the teams for solution establishment, risk, and change management.

The "no excuse for failure" principle implies:

- Involved human and technological resources are fully qualified to deliver the required and documented solution.
- The involved resources are allocated and committed in such a way that their work is done and that the results are delivered without costly interruption, delay, and cost overrun.

Simulated Accept-Testing allows timely and pertinent decisions about required changes to take place without disturbing the progress of complete solution delivery.

The objectives and visions cannot be the strategy on their own because the strategy is also the way, the method chosen to meet the objectives and make the vision come true, that is, the strategic initiatives.

It is very difficult to replace people on teams, especially the people leading or managing the teams, because the people that are asked to leave quite often interpret this as a personal defeat. The people that stay often regard the replacement as a management failure or weakness, which raises their suspicion of further blunders to come.

You can recognize a high-quality strategy by the fact that the results correspond to the objective that has visibly (i.e., documented) been adjusted to what the stakeholders need and expect in the end.

The leaders do not make decisions too often, they make sure that the teams involved with their strategic initiatives are established in such a way that they can operate efficiently and make appropriate decisions by themselves within the scope of the initiative.

2

Team Building for a Strategic Initiative

A Strategic Initiative is taken because an organization or someone with enough power, the sponsor, has seen or been convinced about an opportunity worth going for or a threat that can be avoided or mitigated. This opportunity or threat is the initial cause behind a need for change to a better situation.

The initial challenge of the Strategic Initiative sponsor is to formulate the basic need and the initial scope, and to find the competent people that can and will be involved with this Strategic Initiative to make it successful.

Team building is concerned with the establishment of the best possible organization to perform the Strategic Initiative:

- Selection of people to become key-stakeholders in the Strategic Initiative
- Establishment of the teams of people and the roles and responsibilities of the people in the teams
- Definition of the roles and responsibilities of the teams to perform the tasks required during the lifecycle of the Strategic Initiative
- Establishment of the physical and technological environment within which the chosen people can act and communicate in an optimal way
- Establishment of standards to be used for processes, documentation, and deliverables in order to manage the quality of work, deliverables, and final solution delivered by the teams

Team building is a way to generate synergy; that is, the teams are organized in such a way that the performance of any team is higher than the performance measured as the sum of the team members' individual performances.

The persons to be involved with the Strategic Initiative will contribute in different ways to the success of the initiative tasks:

- Initiate, approve, and govern the Strategic Initiative
- Implement the Strategic Initiative in the corporate strategy

- Coach and facilitate the Strategic Initiative activities
- Plan (develop) the Strategic Initiative
- Develop solution components
- Implement solution components
- Evaluate the Strategic Initiative quality
- Evaluate the Strategic Initiative performance

In order to ensure the best possible contribution to the Strategic Initiative from competent persons and teams, the sponsor provides them with the information, tools, techniques, environments, facilities, and whatever else that is needed to ensure their motivation and ability to perform well according to the "no excuse for failure" principle.

2.1 GET A STRATEGIC INITIATIVE OFF TO A GOOD START

Irrespective of what the Strategic Initiative is dealing with, which types and sizes of organizations are involved, what kind of people are available, etc. there is only one way to get the initiative off to a good start:

> Get the key-stakeholders together and give them no excuse for failure to define the scope of the strategic initiative.

A key-stakeholder is someone with power, knowledge, experience, and competence within the context of the Strategic Initiative that you need to get involved in the initiative in order to make the initiative successful. It is a person who can create or sponsor the development and implementation of some part of the solutions you need. This part of the solution can be infrastructure such as land or public transport and institutions or it can be usage of new technology that can give you important benefits.

Other stakeholders are the classical ones that might benefit or suffer from the Strategic Initiative without being directly involved. You still need to make these stakeholders as happy as possible, which in some cases means "as little unhappy as possible."

You use communication and remuneration to make stakeholders happy, but in order to make the communication and remuneration successful you need to know the stakeholders and their needs and expectations.

This simple stakeholder-based way to get the Strategic Initiative off to a good start poses some important questions to be answered before you can kick off the initiative:

- How do you identify the key-stakeholders?
- How do you communicate with the key-stakeholders?
- What is "no excuse for failure" in your Strategic Initiative?
- How do you get the key-stakeholders motivated for your Strategic Initiative?
- What is the scope of your Strategic Initiative?

There are no simple answers to these questions because the answers are hidden in the heads of quite a few people and some answers might only appear once you start asking questions to potential stakeholders and look into whatever experience material you can find.

I will show how I have handled these questions with some examples and let you judge the pertinence for yourself.

In all the examples, I have had the role as Coach/Facilitator for the Strategic Initiative sponsor; sometimes supported by other Coaches/Facilitators from my company or from other organizations. When I talk about "we" in the following examples, it means the Strategic Initiative sponsor and the Coach/Facilitator.

In rare cases, I have been faced with the problem of making some key-stakeholders "as little unhappy as possible," which also will be explained by examples for you to evaluate.

2.1.1 The Sponsor Role

The sponsor initiates the Strategic Initiative and signs off on the scope as it is established originally and as it is adapted to new conditions and events during the lifecycle of the Strategic Initiative.

The sponsor establishes the Strategy Governance team.

In the case where a sponsor is a group of people such as shareholders or a public institution, such a sponsor is represented by a de facto sponsor role.

The de facto sponsor is a person who performs decisions on behalf of the original sponsor and who participates directly in planning and implementation on a high level, for example, on program management or the corporate management level.

The original sponsor might not exist as a person at all. Imagine a defense budget and a political decision to make the defense green with a part of the defense budget assigned to this objective. In this case, the sponsor is the government and the defense minister, but the government and the defense minister are not personally visible in the Strategic Initiative. However, a person has been made responsible for this task with a budget, and this person is the de facto sponsor.

For the purpose of Strategic Initiative treatment in this book, I will use the term "sponsor" for the de facto sponsor. If reference is made to the original sponsor organization, this is explicitly explained.

It is important to get to know the people with power and influence in the original sponsor organization. These people are key-stakeholders who are kept informed about the Strategic Initiative progress and who continually are motivated to support the initiative and the (de facto) sponsor.

The sponsor might be a CIO, a CEO, a CFO, or simply a program or a project manager appointed for the specific purpose of the Strategic Initiative.

The sponsor person might have more than one role in a Strategic Initiative.

2.1.2 The Coach/Facilitator Role

The Coach/Facilitator delivers the quality system and supports all directly involved team members and managers with appropriate guidelines, procedures, and documentation standards during the length of the Strategic Initiative.

The Coach/Facilitator role is most often delivered out of the corporate project office if this has been established.

The Coach/Facilitator ensures high-quality preparation and conduction of workshops, studies, and working conditions in support of teams to be established or already performing work.

The Coach/Facilitator has an important role to ensure the establishment of "no excuse for failure" teams.

The role that my company most often has delivered to our clients is the Coach/Facilitator role, especially in the case where the client wants to work according to a predefined set of methods.

In most cases, more than one person will take on the role of Coach/Facilitator during a Strategic Initiative because different organizational levels and objectives of teams or workgroups demand different knowledge, skill, and experience from their Coaches/Facilitators.

A Coach/Facilitator who coaches a general manager or a program manager will normally have a profile other than a Coach/Facilitator looking after an agile team of developers and implementers simply because the methods and standards used are very different and therefore demand different experience and knowledge from the Coach/Facilitators.

You can combine a group of Coaches/Facilitators in a Process Governance Team that supports one or more teams and workgroups under a Strategic Initiative.

In most cases, my company has employed competent external Coaches/Facilitators for our own Strategic Initiatives. When being external to the coached/facilitated organization, Coaches/Facilitators can contribute considerably to the value of the result of Strategic Initiatives because of their broader experience and because an external Coach/Facilitator has no constraints of thought and ideas based on former experience with the management of the coached/facilitated organization.

2.1.3 The Unknown Unknowns

Before talking about how the Coach/Facilitator and the Sponsor can work together to identify key-stakeholders and to define the initial scope of the Strategic Initiative, I refer to a press conference with U.S. Secretary of Defense Donald Rumsfeld in NATO*, Brussels, June 6, 2002. Mr. Rumsfeld outlined the challenges faced in a complex defense situation, which could just as well be a Strategic Initiative situation:

> **Question from the audience:** Regarding terrorism and weapons of mass destruction, you said something to the effect that the real situation is worse than the facts show. I wonder if you could tell us what is worse than is generally understood.
>
> **Rumsfeld:** Sure. All of us in this (defense) business read intelligence information. And we read it daily and we think about it and it becomes, in our minds, essentially what exists. And that is wrong. It is not what exists.
>
> I say that because I have had experiences where I have gone back and done a great deal of work and analysis on intelligence information and looked at important countries, target countries, looked at important subject matters with respect to those target countries and asked, probed deeper and deeper

---

* Donald Rumsfeld press conference, June 6, 2002. http://www.nato.int/docu/speech/2002/s020606g.htm (With permission of NATO).

and kept probing until I found out what it is we knew, and when we learned it, and when it actually had existed.

And I found that, not to my surprise, but I think anytime you look at it that way what you find is that there are very important pieces of intelligence information that countries, that spend a lot of money, and a lot of time with a lot of wonderful people trying to learn more about what's going on in the world, did not know some significant event for two years after it happened, for four years after it happened, for six years after it happened, in some cases 11 and 12 and 13 years after it happened.

Now what is the message there? The message is that there are no "knowns." There are things we know that we know. There are known unknowns. That is to say, there are things that we now know we do not know. However, there are also unknown unknowns. There are things we do not know we do not know. So when we do the best we can and we pull all this information together, and we then say well that's basically what we see as the situation, that is really only the known knowns and the known unknowns. And each year, we discover a few more of those unknown unknowns.

It sounds like a riddle. It is not a riddle. It is a very serious, important matter.

There is another way to phrase that and that is that the absence of evidence is not evidence of absence. It is basically saying the same thing in a different way. Simply because you do not have evidence that something exists does not mean that you have evidence that it does not exist. And yet almost always, when we make our threat assessments, when we look at the world, we end up basing it on the first two pieces of that puzzle, rather than all three.

Together with the key-stakeholders to be involved in the Strategic Initiative, we want to define the Strategic Initiative scope precisely:

- Why the Strategic Initiative is required
- The organizations to be involved and why
- Solutions and products to be delivered and why
- The needed quality of the solutions and product and why

We are faced with conditions that are not only the known ones. The Strategic Initiative conditions also comprise the unknown ones that we will meet in the future.

The current and future conditions present us with threats and opportunities that demand our response:

The known knowns	These conditions have been documented in whatever requirements spec or problem list that has already been established.
The known unknowns	These conditions are known by stakeholders that have not yet been involved. It is the current pertinent tacit knowledge that you need to activate to understand more about your opportunities and threats.
The unknown knowns	These conditions should have been documented in requirements spec and problem lists, but these conditions are so obvious that no one thought about documenting them.
The unknown unknowns	These conditions you might discover as group synergy or by simple luck, but you will not find them if the minds of you and your stakeholders have not been set to be observant and creative—to "think out of the box."

2.1.4 Stakeholder Identification

Strategic Initiatives can have many different preconditions that will play an important role for the initial identification and selection of stakeholders:

- There might be a requirements specification that explains in detail what is expected from the Strategic Initiative.
- We might have signed a contract that explains in detail what is expected from the Strategic Initiative and what our roles are expected to be.
- There might be only a wish list established by the original sponsor who has only a vague idea about what is at stake in order to succeed with the initiative.
- There might be only a list of problems to be resolved and it is up to us to decide on our respective roles (Coach/Facilitator and Sponsor) and the roles of other key-stakeholders to be involved in the Strategic Initiative.

Once contracts, direct orders, or other agreements have ensured that the original sponsor supports the initiative, we start searching for the key-stakeholders to participate in the future Strategy Governance Team.

Although the preconditions are important and we have to know and understand them, they are historical. Requirements specifications, wish lists, and even contracts are merely guidelines to get the initiative started off.

In most cases, you will find a first answer to why the Strategic Initiative is conducted in these preconditions, but sometimes even this important "why" is answered only with a political decision that in vague terms addresses the real opportunities of the Strategic Initiative.

It becomes a task of the chosen key-stakeholders and ourselves to answer the why question and the what and the when and the how questions as well of course in order to fully get to a common understanding and agreement about our opportunities and threats.

If we do not get the why right relatively early, we will have a hard time to motivate future key-stakeholders to get on board and to stay active in and motivated for the Strategic Initiative until the initiative closes out. Only if people feel that they contribute to something valuable can you keep them motivated. We keep this feeling alive by involving the stakeholders in Strategic Initiative processes where they can and will contribute positively and visibly to the result.

The Sponsor has knowledge about who the key-stakeholders might be and the Coach/Facilitator has knowledge and experience about how stakeholders can be treated and made happy, once identified.

To help us identify the key-stakeholders of the Strategic Initiative we establish the organogram with the organizational units to be involved (Figure 2.1)

We will also use the Value Chain to identify pertinent organization units and business processes for the Strategic Initiative (Figure 2.2).

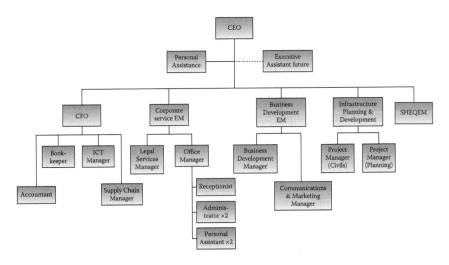

FIGURE 2.1
Organogram.

FIGURE 2.2
Value chain (from http://bettyfeng.us) and strategy.

For each pertinent organization unit we outline their quality objects:

- Products
  - What, how, and why they deliver what to or receive what from other organizational units externally and internally
  - Quality, that is, why the clients like what they deliver
  - What they produce in support of their own processes
  - Key figures such as number of employees, annual production volume, cost of operation, etc.
- Processes
  - What they do and why they do it
  - Productivity measures
  - Efficiency
- Organization
  - Functional Areas
  - Managers
  - Clients internally and externally
  - Key knowledge persons (skill, experience, competence)
  - Communication and knowledge sharing

We discuss and document why each of these quality objects has an influence on or is the reason behind the known problems and objectives for the Strategic Initiative. We use this knowledge—and, quite often, the lack of knowledge on our part—to prepare ourselves for the initial dialogues with identified key-stakeholders.

Sometimes enterprise management or another original sponsor has already pointed out the key-stakeholders, but we still need to get to know their motivation for involvement in the Strategic Initiative.

We ask the key-stakeholders to tell us what they think the opportunities and threats are if we pursue the Strategic Initiative. We also ask them what they think the opportunities and threats are if we do not pursue the Strategic Initiative.

Finally, we ask questions that are more personal in order to measure their motivation for getting involved with the Strategic Initiative:

- What is your vision of the future situation when the Strategic Initiative has been successfully completed?
- What could your mission and role be in this Strategic Initiative if or when you get involved?

This initial dialogue with potential key-stakeholders who are potential participants in the Strategy Governance Team makes it possible for us to answer the why question with focus on real pertinent opportunities and threats. We obtain the information we need to prepare the Process Quality Assurance (PQA) to kick off the Strategic Initiative on corporate or strategy governance level.

You use the same type of dialogue to prepare PQA processes for other PQA Teams that are established on a lower level.

2.2 KEY-STAKEHOLDER SELECTION EXAMPLES

The examples shown have all been used to prepare PQA workshops or similar team building brainstorming workshops to kick off important Strategic Initiatives.

The first two examples comprise comprehensive changes to business processes required because the market and the client needs had changed with availability of new technological opportunities.

The third example is the private bank information system swap that was provoked by its technology being out of date. Contrary to the two former cases, this one has a demand for as little change as possible to business processes. It has to be proven that the new COTS application can at least handle the known business processes when it has been set up as required by the users.

2.2.1 A Merged ICT Consulting Enterprise Project Management Improvement

In this case, my company was called in as Coaches/Facilitators based on our long-time experience of program and project management of Information System implementation in international organizations.

The client had fully trained and certified program and project managers based on project experience in the Information and Communication Technology (ICT) industry, but also from implementation of SAP-based information system solutions.

SAP is an Enterprise Resource Planning COTS software package well known worldwide that can support business functions in any type of organization. The original application modules comprised Production Management, Inventory Management, Distribution Management, and Financial Management.

On the experience and competence level we were peers, but with very different and complementary backgrounds.

2.2.1.1 Initial Problems and Needs

The ICT and IT consulting firm is a merger of three organizations. This merger belongs to an international ICT consulting corporation with more than 30,000 employees. The local consulting firm runs numerous projects and smaller tasks to implement often-comprehensive client solutions and to implement their own Information System solutions based on SAP and other integrated software.

Project Management is a primary skill and competence of the ICT management consultants. In the various departments and functional areas there are different traditions for handling of projects and for registering the time spent on these.

It is essential that the time spent on all projects be recorded systematically and reliably, so that the experience can be used for future client offers and for improvement of the procedures used internally and externally. Some of the recorded time is used directly as a basis for billing of customers.

Registration of time spent must be coordinated so that the principles are the same from functional area to functional area, but without the specific requirements of each functional area being ignored. The different functional areas use specific Work Breakdown Structures with specific phases and milestones in support of their specific project planning and tracking needs that all must be supported by the new cross-organizational solution.

The Strategic Initiative must establish the basic requirements for future registration of time spent and develop an implementation plan, so this initial time registration can be started in April.

Since the time spent must be recorded in relation to projects, a very simple way to visualize the projects must also be conceived so that employees can record their time spent.

The Information System to be implemented will use a COTS application that includes the following needed functionality:

- Central repository of projects, resources, teams, user rights, and fundamental standards.
- Scheduler for the creation of projects, allocation of resources (from an assigned team) on tasks, and planning and follow-up on phases and activities.
- Time recording for employees to record time spent as resources on the created projects.

The project group's (this was the Strategy Governance Team) initial objective and task is to establish the scope and the initial requirements for the future Project Management Information System that meets the requirements for time recording, which can be transferred to SAP when it has been checked and approved.

2.2.1.2 Sponsor and Strategy Governance Team Member Selection

The sponsor on this Strategic Initiative was the Project Support Office Manager. The project had attention from and the budget approved by the corporate HQ because the same need was recognized worldwide. The corporate HQ was the original sponsor.

The sponsor and I selected the team members among the following stakeholders:

Managers with long-time experience from conducting all types of ICT projects

Managers with experience from internal and external SAP projects

Project office managers to support the future solution implementation

Education manager to support the future solution implementation

Finance support manager in charge of internal accounting and invoicing of clients

The original sponsors (corporate stakeholders) were waiting for the result, but they were not directly involved.

After successful implementation of the new Project Management solution, my sponsor was promoted to implement the solution across Europe.

2.2.1.3 *Strategy Governance Team Members*

Network Integration Manager
Managed Service Manager
Enterprise System Integration Manager
Business Solutions Manager
Finance Manager
Finance Support (SAP)
Education Manager (internal education)
Education Manager (client education)
Project Support Office Manager (Sponsor)
Project Support Office PM Support
Project Support Office SAP Support

The team size of 11 people is not optimal (we regard 4 to 8 people as optimal) because too many ideas of marginal interest make the workshop take longer time to conduct without improving the result. You may reach a high number of participants in the Strategy Governance Team when many cross-organizational business procedure changes are required.

In this case, the Strategy Governance Team was big because the participants from different merged organizations needed to find a common playground for future Project Management, which would imply quite a few business process changes even in light of that part of the objective, which was to respect the individual behaviors and needs as much as possible.

2.2.2 Web Catalog Factory (WCAT) Order Management Improvement

In this case, my company was called in to develop and implement its future Project Management solution to be fully integrated with its Order and Production Management systems.

The case involves usage of the PQA procedure to establish motivation and involvement across all departments, which will be explained further under PQA examples.

Our own technicians and developers would work closely together with their own IT staff in order to develop the integrated solution and to prepare the education of the staff that would use the implemented information system solution.

2.2.2.1 Initial Problems and Needs

The company is one of the world's most important suppliers of systems for production and maintenance of web-based catalogs to be reachable from many media. The catalogs can retrieve information across many databases and data media from alternative data providers.

Customers are typically large international companies with internationally distributed production and sale of their products.

Projects comprise internal development of standard systems and tailor-made customer solutions. Client solutions may be recurring orders involving only a few adjustments in each case, or they may be completely new orders with varying degrees of solution development.

Customer projects today require no great intensity in cooperation with the customer, but there is a tendency for it to become increasingly necessary to involve the client's employees in system development, especially when it comes to more advanced user interfaces.

The wish for a new and better project environment must be seen as a natural progression toward greater efficiency and steadily improved competitiveness and customer service.

To improve efficiency, a better overview of production processes and their contexts is needed. Production process data must be collected systematically in a way that allows for preparation of standard times for standardized procedures. Production process data must show how cooperation between the various departments is handled during a project so that handover of partial deliveries can be ensured in time with the right quality.

It must be possible to assign available scarce resources across national borders.

It must be possible to prioritize projects, so that less critical projects do not inadvertently get critical resources assigned at a time when a more critical project requires their usage.

In order to increase competitiveness, besides through increased efficiency, it must be possible to analyze the impact on the production capacity of each customer order accurately in order to be able to build a realistic plan for cooperation with the customer on completion of a delivery.

The teams that are expected to complete a project must be shown as early as possible in the project database so that departments can plan their resource availability. The overview of resource capacity and the status of ongoing projects must be internationally available to sales and project management.

The future project information system must create visibility of active and planned projects so that the cooperation between project participants, project managers, and department managers can be handled on a realistic basis.

Acquired experience must be classified and available so that it can be used for future estimation and projects planning.

It is especially desirable that project participants can visualize that they have been good at planning their projects.

The Project Management Information System combined with improved business processes must ensure reliable and consistent information and communication in support of the daily project management and the long-term capacity planning of resources.

It is essential that the organization is prepared to exploit the system functionality effectively before the system is fully implemented. Many small victories are better than one "big bang."

It is important that workflows around the system usage are in place and that there is education covering the concurrent usage of all integrated information systems for project management.

Finally, it is important that pertinent standards are established and implemented before the systems go into operation.

2.2.2.2 *Sponsor and Strategy Governance Team Member Selection*

My sponsor on this project was the General Manager. Besides getting the new information system in production as fast as possible, his primary objective was to establish broad motivation for the implementation of the future solution. It was therefore essential that all future types of users were represented in the Strategy Governance Team.

The following departments were identified to be involved:

- Administration
- Sales
- Client Support
- Planning
- Development
- Production
- Quality
- IT

2.2.2.3 *Strategy Governance Team Members*

The following were Strategy Governance Team Members:

General Manager
Deputy General Manager
Development Manager
Finance Manager
Sales Manager
Client Service Manager
Client Project Manager
Quality Manager
Production Manager
Deputy Production Manager
IT Development Manager
Lyngso coach/facilitator (Project Management Information System Development and User Education)

The Project Governance Team is rather big because a large number of business procedures will change, which needs involvement in decision-making across the whole organization. No single business function has the ultimate recipe for the future workflow.

2.2.3 Private Bank Information Systems Swap

In this case, I was contracted to be Coach/Facilitator for the Program Manager for a huge information system swap program involving:

- Swap of technology
- Swap of databases
- Complete workflow requirements documentation
- A big COTS back office application bought by top management yet to be set up and implemented
- Integration of a COTS-based reconciliation system
- Integration with a COTS-based fund management system
- Procurement and implementation of their future COTS-based finance solution
- Development of client and risk reporting
- Development of system integration
- Procurement of competent resources

Initially my title was Project Manager with direct report to the Program Manager. The initial objective was to get the program projects under control and to establish reliable progress tracking. In this respect, I was the Coach/Facilitator for the Program Manager, but that title was not recognized in the private bank organization.

2.2.3.1 *Initial Problems and Needs*

We have already seen some of the preconditions of this big program, but there are even more alarming reasons to perform immediate recovery:

- There was no Program Governance Team in place.
- The former IT Director had left the bank.
- Projects were only informally established without visible management.
- There were no requirements specifications.
- The current IT business system technology had to be swapped out quickly because it was without original vendor support (spare parts were only available from faced out equipment that was difficult to find).
- Project activities were performed by internal and external consultants in IT, the external ones invoicing monthly, but no one understands why the activities are performed, what the result is, and how the activities are performed.

- Training of busy future users had been initiated using the COTS application without adaptation to the users information system needs.
- The future users of the COTS application were very dissatisfied with what they saw during training and demonstrations.
- There was general frustration with the program progress, especially on the management level.
- A Program Manager had lost control.
- There was a new IT Director who did not like to see his budget spent on useless deliverables.
- There was no time to waste as the bank was faced with a risk of total IT breakdown every day.

I was not employed directly as the Coach/Facilitator for the IT Director initially; he only later became my Sponsor. My contract to be Project Manager with reference to the Program Manager was signed by the Program Manager.

The IT Director was new in the organization and he did not know anyone other than the Deputy General Manager who had signed his own contract. He was of course aware of the dissatisfaction and management frustration with the COTS implementation.

The COTS implementation Program Manager and the predecessor to the new IT Director had signed the contract that allowed external IT management consultants to send monthly invoices against the IT budget based on worked hours, not on results.

My first task was to get to know the Program Manager and to dig into what was happening in the program.

We quickly discovered that the external and internal consultants were doing GAP analysis. They evaluated all existing software functionality in order to establish whether it could be handled by the future COTS-based solution.

The internal and external IT management consultants involved with the GAP analysis rarely talked to the business users unless they had been involved directly with the former solution development.

There was no formal project plan and the documents from the GAP analysis were unusable because they just stated whether a solution could survive, never why or how it was accepted or refused.

Many future COTS application users had been given training in the COTS application that had not been adapted to their needs (yet). These needs were not documented. All users talked openly about their frustration to the new IT Director.

I presented these problems to the Program Manager, my contractual sponsor. As the Program Manager directly employed me, I was of course loyal to him and his decisions, so he was the first person to know about my observations that were of no surprise to him.

In full agreement with the new IT Director, we all agreed that we could not start criticizing the active program because:

- The program was initiated and backed up from top management level.
- Criticism this late would create more frustration in the program team, which was already under heavy pressure.

We decided that I would present myself and a recovery plan ("Suggested Improvements to the Project Plan") to the current Program Manager and his managing team members.

The Program Team was established with:

- Primarily IT-based resources
- A few selected business operation staff with IT solution knowledge
- One COTS vendor representative
- The manager of the external management consultants

There was absolutely no finger pointing at anyone in my presentation.

The presentation had no suggestions about enhanced management, only suggested changes to activities and objectives to speed up progress and get real results.

The Project Manager presentation meeting resulted in the following minutes that were distributed to the Program Management Team and the IT Director. The Program Manager and the IT Director informed general management and bank departmental management about the situation in a meeting. The Program Manager criticized me heavily in this meeting, to which I was not invited.

When you read the minutes you will understand the reaction from the Program Manager, and you will probably find a way to be more diplomatic yourself. However, it all depends on the situation you are facing, and in this case, there was absolutely no time to waste. What I did not know was that general management had already decided to replace the Program Manager.

Here are the minutes of my presentation to the Program Team that presented a way forward to be accepted by the IT Director (it was his budget) and the Program Manager (who agreed fully to the suggested project improvements):

> The key to our success is that we can make the COTS application handle the Business Procedures that today are supported by outdated information systems running on non-supported technology and that we can build interfaces to several applications already active and currently interfaced with the outdated information systems. The delivered result must support all business functions ideally 100%, but realistically at least on a level where the banking operations are legally compliant.
>
> We can succeed by simulating the real bank operations in parallel and interactively on the COTS application and on all other systems to be used in the future solution, which will demand and also ensure that the necessary communication between business competence areas of the bank and the bank's IT side is established.
>
> The target for this work is to establish an end-to-end test model, where we simulate all bank transactions with a variation ensuring that basically all possible business variance is tested.
>
> This test model can be used again and again—also for testing of other applications; and it will be used during migration of systems and modules relating to the Core Banking Information System and the Corporate Risk Management Systems in order to verify that all system integration produce the required results.
>
> The test model will focus on bank procedure functionality, not on COTS application functionality. The COTS application functionality will only be tested under Accept-Testing after it has been set up by experienced technicians to comply with the Banking Workflow requirements to be documented by banking user staff and departmental management.
>
> To produce the business function test cases for our (acceptance) test model we need to reorganize the project organization into Workgroups with the following capabilities:
>
> - Make necessary decisions
> - Initiate required work
> - Perform the work
> - Evaluate the work done

Especially the capability to make and communicate decisions is important. This does not mean that all persons (roles) have to work together concurrently all the time, but it does mean that all roles and lines of communication are precisely defined before work is initiated in a Workgroup. This will avoid any waste of time waiting for decisions to be made or waiting for facilities to be available.

Workgroups will cover both business functionality and the IT environment.

Assigned resources will typically work in more than one Workgroup, which will contribute to facilitate the communication between Workgroups.

Each Workgroup will have the responsibility to produce test cases (Workflows) of the production of specific (business) products (transactions, corporate actions, static data, reports, etc.) used either directly by bank clients or by other internal or external organizations/functions.

The consequences for the activities in current workshops are the following:

- Workshops will be used only for production of test cases (Workflows) to be used by the COTS system engineers for the setting up of supporting functionality in the COTS application.
- More than one Workgroup can participate in a workshop in order to ensure that all necessary knowledge and experience is used for production and evaluation of the produced test cases (validity, relevance, completeness, and consistency).
- The test cases are not developed in detail on the workshops—each individual Workgroup does the actual definition of test data within an agreed deadline.
- Each Workgroup participant will have access to use a current information system playground version for implementation, documentation of current Workflows, and for desktop-based verification of the results.
- Once the COTS application has been set up to accommodate a business function in its own playground IT environment, the relevant Workgroups will start producing training material to be used during Accept-Testing and later—after approval—for training of business end users and IT support.

The test cases will later on be used in formal Acceptance Test Sessions (workshops), when a test ready version of the COTS application has been set up by competent resources (not yet assigned to the program) to be Acceptance Tested.

In parallel with Workgroups producing business test cases, other Workgroups from IT operation, IT development, and IT infrastructure will be active:

- They prepare, operate, and maintain an agreed level of system service for current applications and later on the COTS application in a playground to be established.
- They set up the COTS application to be ready for bank user Accept-Testing and later on for user training and training of IT support personnel.

- They set up the COTS application and integrated information systems IT environment with backup, rollback, job creation, job execution, interface setup and timing, controlled operation, etc.

The idea is to generate a complete overview of and control with the capabilities of the COTS application-based Core Banking solution ready for Accept-Testing once all at this point known preconditions are met. We know that the COTS application quality will surprise us a lot during setup, e.g., the current list of problems and issues is already too long.

This primary target will be met by the end of June.

Between the end of June and until the end of September we will build (and simulated acceptance test) all the bits and pieces required for interfaces from current applications to the COTS application and for database conversion from the current information systems to the COTS application SQL database.

The objective is to replace the outdated information systems completely with the COTS application enhanced with functionality of interfaced known systems, and new information systems, e.g., the future finance solution to be established into the final new Core Banking solution.

All developed elements are documented:

- Test model
- Business processes
- User interfaces
- Jobs
- Interface solutions
- Data conversion rules and routines
- COTS application setup
- Reporting requirements and solutions

For each element it is declared why it has been done and why it has been made the way it is—hereby ensuring full traceability of issues and problems to simplify future maintenance, integration, and migration.

Project activities and decisions are documented with planned outcome, roles, responsibilities, and achieved results.

The project progress is documented in status reports with lists of outstanding issues. This work is handled under the heading of "Project Administration," which will also look after project accounting and cost control.

All formal tests are documented with errors and issues logged from identification until final resolution.

The document standards are established and await agreement from the Core Banking Project organization (starts tomorrow).

All results will be documented immediately as outlined above (starts tomorrow).

This means that the work results achieved until today must be documented with the quality outlined above, if at all relevant (starts after the end of the current workshop).

All further workshops are rescheduled after today in order to allow for reorganization and completeness of Workgroups, i.e., expansion of the current user participation to be agreed with departmental management immediately.

Two days after this workshop, the Program Manager was replaced with me by general management (the original sponsor). I tried to warn my predecessor about the risk of this event, but the General Manager had already made this decision with immediate execution and without asking permission from the IT Director.

2.2.3.2 Sponsor and Strategy Governance Team Member Selection

As the former work had created more frustration than solutions and as no functional results, not even the COTS application in a technically accepted state, had been delivered, my new Sponsor, the IT Director, and I were ready for a new start, and so was the Program Team with their new Program Manager.

The first action performed by the IT Director backed up fully by General Management was to release all external management consultants. We only kept two technical consultants with COTS application expertise.

We then called a meeting where we invited two other key-stakeholders for the "new" program:

- The deputy General Manager
- My deputy Program Manager inherited from the former Program Manager

Before the meeting, I had used my research results to draw up my dream Program Management organization that would involve all business units including IT actively in Workgroups with agreed objectives and deliverables.

The suggested new Program Organization is shown in Figure 2.3.

The way that we ensured IT, Sub-contractor, and Vendor ownership of the technical solution components delivered to IT and Business Workgroups cannot be shown directly in the organogram.

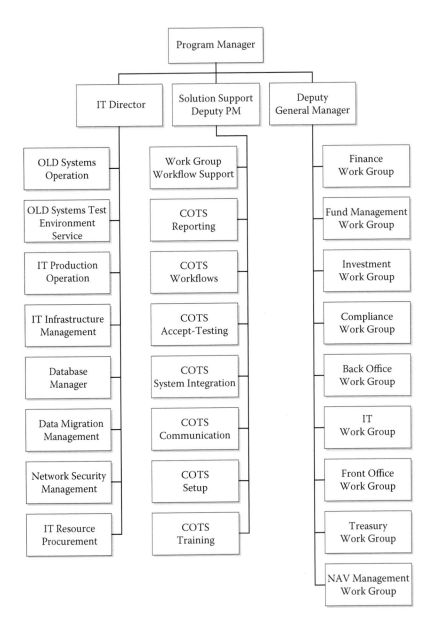

FIGURE 2.3
Program organization example.

This ownership commitment was based on contractual conditions that were developed especially to this end. These contractual conditions are explained in Chapter 4 under Procurement.

The principles used to establish the new Program Management organization were:

- Establishment of the Strategy Governance team.
- Involvement of competent resources in organizational elements that get no excuse for failure, for example, the Program Manager ensures access to required document standards to be agreed by and used by all involved Workgroups. Many organization elements had no resources at all which was a problem and an opportunity at the same time because it allowed the new Strategy Governance Team to establish these elements in an optimal way.
- Business including IT ownership of all solution components to be delivered.
- Thorough quality management ensured by the Strategy Governance Team and directly managed by the Program Manger and his deputy. The deputy Program Manager had several years of experience from all business operations and could evaluate and challenge the quality of all Workgroup Workflow documents delivered for COTS application setup.
- Although some business function managers were more important than others were, it was clearly stated that Program Workgroup involvement respected only program specific management appointments. All Workgroup Managers reported to the Strategy Governance Team. Most of the Workgroup managers were also Business Organization managers with responsibility for the business processes treated by the Workgroups, which made it easier to ensure that the best possible resources got the necessary priority to perform the work with Workflow documentation.

PQA workshops were deemed unnecessary because only one key success factor was pertinent:

All business functionality must be compliantly supported by the new COTS application and other interfaced applications before the old technology breaks down.

To fulfill this success factor we had to execute all the activity in the project plans to be developed. The Strategy Governance Team possessed together all the knowledge, skill, experience, power, and competence necessary to establish the project plans with deliverables, solution components, organization, communication requirements, etc.

We replaced the PQA Workshop with a special Information Requirements Study (IRS) workshop, where all departmental managers and their deputy managers participated (20 people in all). In the IRS workshop we established all business process Workgroups and their scope of requirements specification Workflow production.

The project plans of course were fully documented and signed off by all pertinent managers for execution. All activities within the scope were fully documented and approved by the relevant Workgroup managers as if a PQA process had been performed.

The procurement of resources and integrated system solutions was an important part of the project plan that you normally only find in such big programs. The IT Director coached by me managed the procurement. My company delivered the concepts for requirements specifications and contractor contracts.

The final organization had more than 100 participants of which 50% were employed full time during the implementation of this Strategic Initiative that finally was run as a program.

2.2.3.3 Strategy Governance Team Members

The following were Strategy Governance Team Members:

> Deputy General Manager
> IT Director
> Program Manger
> Deputy Program Manager

The Strategy Governance Team was small because no cross-organizational business procedures were allowed to change unless absolutely necessary for legal or required business change purposes. If major procedural changes need implementation later, the Strategy Governance Team can be expanded.

In this context, completeness of the program with respect to projects and resources can be decided with support from the group of experts already known by the Strategy Governance Team.

A formal PQA process was not needed. The documentation that would have been established through the PQA processes still had to be produced in order to document the complete program plan and project plans that in the end delivered the required result. This documentation comprised detailed agreed upon requirements specifications for the deliverables from all Workgroups.

The deputy General Manger proved to be very efficient whenever conflicting interests surfaced between business operation, IT operation, and Program Management. Especially where business departments thought they could have just an advisory role, he was very convincing in his demand for direct involvement. On the other hand, we never succeeded to involve the finance department and the risk management department in a responsible way in the program. Both of these departments got general management support in running their own parallel projects, which created a lot of problems and delays, but did not prevent the program from succeeding in the end. Both of the managers from these departments later left the private bank.

2.3 ORGANIZATIONS FOR STRATEGIC INITIATIVE SUPPORT

The typical organization constructs involved with Strategic Initiatives from initial establishment to final implementation and governance are:

- A Project Office established in the line organization in support of all projects and programs in the corporation.
- A Program Office established as a secretarial function and an executive organization representing a Strategic Initiative Governance Team.
- Decision-making and executing teams established for the development and implementation of Strategic Initiative results under continuously changing conditions and risk.

Some generic types of organizations and teams are outlined next.

2.3.1 The Project Office

The Project Office is established on a corporate or enterprise level to support all teams that run projects in the corporate portfolio of projects, as well as the projects that belong to Strategic Initiatives or programs.

The Project Office has the role to support efficient communication among teams and between teams and corporate management and other stakeholders.

The Project Office governs the corporate methods, tools, and techniques used in the context of Project Management such as the corporate Project Management Information System.

This allows the Project Office to ensure valid and appropriate progress information from and to all teams and to and from the teams working under active programs and Strategic Initiatives.

The Project Office collects the planning and tracking information from all projects and ensures its completeness and value based on communication with all executing projects, as well as the projects established under programs.

This planning and tracking information is made available to all authorized persons.

Access to and distribution of planning and tracking information is also supported by the corporate Information Management System that contains and controls much more information than the project and program related one, for example, business performance measurement, financial information, client service information, etc.

The corporate Project Management Information System is normally integrated with the corporate Information Management System, but not always.

2.3.2 The Program Office

The Program Office is a secretarial facility and knowledge management center for a specific program.

It is composed of people with in-depth knowledge and experience from technical, legal, and political environments that are important knowledge areas for the program.

Based on this knowledge and experience, the Program Office can ensure that important information about rules, technological development, and political wishes is made available to the Program Governance Team.

The Program Office does not make decisions, but it prepares the information required for decision making by the Program Governance Team. In this respect, the Program Office can demand progress information from the Workgroups and the business organizations under the program.

Most often, the progress information requested by the Program Office has been prepared and validated technically by the Workgroup teams in question supported by the Project Office.

2.4 TEAM TYPES FOR STRATEGIC INITIATIVES

Our primary principle for establishment of teams is our agile principle:

> The best solution architectures, requirements, and designs emerge from self-organizing teams. Self-organizing does not ensure this situation; it is the fact that the teams are established in such a way that self-organizing is possible that ensures the situation.

For efficient Strategic Initiative handling, we have used different team types:

- Some teams are more dedicated to establishing teams and providing the teams with "no excuse for failure." They are good at defining scopes and setting targets. They are excellent at producing requirements specifications and setting up contract terms. They are excellent at communication with all types of stakeholders.
- Some teams are dedicated to producing and implementing solution components. They are good at responding to requirements and business Workflows with solution components. They have the skills and competences needed to select and utilize the best possible technology in the production of solution components.
- Some teams are dedicated to integration, Accept-Testing, and solution implementation. They are good at translating requirements specifications into business Workflows. They have business skills and competence that allow them to ensure legal compliance and secure operation and usage of integrated solution components. They can train and coach solution users.

The different types of teams have one set of capabilities in common (Figure 2.4).

We ensure that the teams can make decisions, initiate work, do the work, and evaluate the progress and the results obtained for process improvement, that is, perform their own quality management.

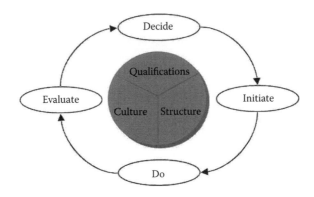

FIGURE 2.4
Team capabilities.

At regular intervals, the teams evaluate how to become more effective, then tune and adjust their behavior accordingly. The evaluation comprises communication between teams and with their environment of stakeholders.

This team capability to perform continuous quality improvement is ensured by:

- Dynamically providing the teams with resources that have the needed competences.
- Ensuring that an appropriate culture is in place for efficient communication.
- Ensuring that the required systems, tools, and technology are available to the team.

This is part of the "no excuse for failure" team principle.

Teams that miss some key competence might inadvertently have been established for two distinct reasons:

1. The scope of the work for the team is defined too large, which makes it impossible to combine the required skills and competencies within one team.
2. The team has been allowed to or been ordered to initiate work before all required skills and competences have been defined and allocated to the team.

In these cases, you will have to add time for coordination and communication across organizations or with external parties that might not have the same priorities or attitudes as your team. This adds duration, potential conflicts, bad results, and other negative effects for which you have to plan and to which you must respond.

By using appropriate PQA such as outlined in the introduction and discussed in detail in Chapter 3, these team-building problems and risks can be completely avoided.

In most Strategic Initiatives, you will work with more than one team, where each team is built according to the rule discussed previously. The cooperation between these teams is ensured with efficient communication and coordination managed by Program and Project Managers and supported by the Coach/Facilitators and the Project Office.

The interrelationship between and the mutual responsibility among the teams is documented in the agreed project plans, the program foundation, or in a contract.

Every time we have been faced with a new organization and a new corporate culture, we have had to rethink our approach to team responsibilities such as discussed in the different case histories, but the basic agile principles have always proven their value (Figure 2.5).

The different team constructs with which we have most often worked under Strategic Initiatives comprise:

- The Strategy Governance Team initiates a Strategic Initiative based on leadership decisions. The Strategy Governance Team ensures the successful conduct of the Strategic Initiative based on efficient communication. It is the top level of Change Management.
- The Process Governance Team coaches one or more PQA teams and is the second level of Change Management. Only Process Governance Teams can propose Strategic Initiative changes to the Strategy Governance Team.
- The PQA Teams lead, manage, and plan activity in order to deliver agreed tangible and measurable results. The Strategy Governance Team is a PQA Team concerned with initiation and kick-off of the Strategic Initiative and is concerned with major changes to the Strategic Initiative proposed by the Process Governance Teams. The PQA Teams can lead one or more subordinated PQA Teams and manage Workgroup Teams. The PQA Teams are the first level of

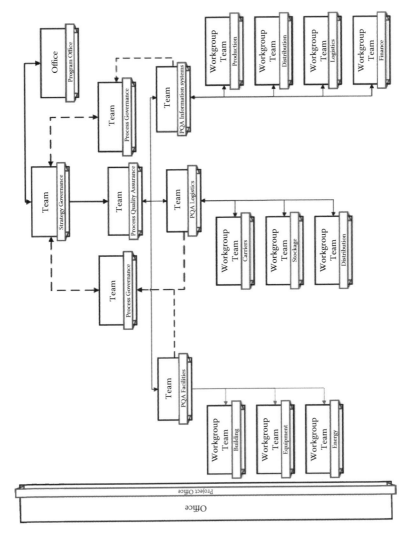

FIGURE 2.5
Strategy Governance Team structure.

change management. PQA Teams can suggest changes to the Process Governance Team that looks after the PQA Team.

- The Workgroup Teams perform production and implementation of agreed solution components. They control the quality of delivered solution components. They report progress in the Project Management Information System that has been implemented by the Project Office. Problems, Risk Conditions, and Events are reported to the PQA Team that manages the Workgroup Team.

2.4.1 The Strategy Governance Team

The group of people that establishes and governs a Strategic Initiative in the enterprise is called the Strategy Governance Team. The enterprise can be any type or combination of private, for profit, governmental, associative, or non-profit organization. The Strategy Governance Team can be from one person to many.

On a level decided by the Strategy Governance Team, it defines the complete set of solutions, processes, and organizations of the Strategic Initiative, that is, the detailed scope of the initiative.

The Strategy Governance Team has the power and the authority and under some constructs the full responsibility to lead the enterprise and to ensure that the enterprise is on track to achieve its agreed objectives.

Important Strategy Governance Team qualities include the following:

- Leadership—because this team establishes the original targets and objectives.
- Power—because this team sponsors the Strategic Initiative and makes the ultimate decisions concerning the direction and the budget of the Strategic Initiative work.
- Authority—because if the authority is not with this team, then subordinate process governance teams and even Workgroups will challenge decisions made and create unwanted conflict.
- Stakeholder Trust—because Workgroups and subordinate process governance teams must be able to work and progress with the certainty of the Strategy Governance Team backing and support; and because if some key-stakeholder loses confidence in the Strategy Governance Team, serious destructive conflicts will surface.
- Responsibility—because Workgroups and subordinate process governance teams will lose confidence and respect if management does

not assume its responsibility, which leads to endless discussions and destructive conflicts.

- Empathy—because even the best formulated critique and demand for change must be investigated in light of personal attitudes among team members and key-stakeholders.
- Good coaching—because in order to be challenged to lead and to let other teams manage and work, good coaching is needed.
- Good facilitation—because in order to cater to good decision-making and team-building processes, facilitation is needed.

The Strategy Governance Team consists of managers with different functional backgrounds:

- Board Members
- Executive Management
- Technology and production
- IT and Information Systems
- Logistics
- Finance
- Client Relationship Management

The Strategy Governance Team leads, guides, or inspires other teams to destinations they would not go to alone by:

- Ensuring external efficiency
- Doing the right things
- Knowing why
- Managing people
- Establishing visions
- Finding opportunities
- Empowering people and teams
- Building the teams
- Having contacts
- Cooperating

The leadership role does not imply that managers of other teams cannot be members of the Strategy Governance Team. Leadership is not a personal matter but a matter of activating knowledge, experience, and motivation across a well-selected group of people with a common purpose.

The Strategy Governance Team establishes the Strategic Initiative based on a thorough SWOT (Strength, Weakness, Opportunity, Threat) analyses that if well done ensures that the why is answered for the initiative.

- The Strategy Governance Team evaluates the strength and the weaknesses of the enterprise in the market and the environment where it will operate; sometimes backed up by industry experts and analysts.
- The Strategy Governance Team evaluates the opportunities for doing business and the threats facing the enterprise doing this business. These threats and opportunities are the key to identification of possible strategic initiatives that can improve the situation of the enterprise.

The Strategy Governance Team has typically assigned my company to coach and facilitate Strategic Initiative implementation after that pertinence has been established with a SWOT analysis.

Only in rare cases have we met the industry experts and analysts employed by the Strategy Governance Team to do the SWOT analysis. These industry experts and analysts produce their own type of reports and give recommendations on the direction to take by corporate management, for example:

A major container line operator used a worldwide-recognized consultancy firm to evaluate the future development of the container line transportation industry and market. The container line operator was facing a need to invest in new capacity of three big carriers, a high-risk multibillion dollar investment, if they should continue to compete in the market.

The management consultancy firm would not recommend a continuation of the container line operator in the very competitive container market.

My company became the strategy implementation coach and facilitator based on continued container line business in this enterprise. We therefore took the container market continuation and the market knowledge of the major container line operator as a given. They were, at that point in time, one of the most important operators in the market.

Our Information Requirements Study analysis and recommendations were based on the business procedures and the needs of the container line clients and the other market operators in continued activity.

We recommended a Strategic Initiative that would improve the competitiveness of the container line.

Our recommended strategy would bring the container line operator in closer contact with the clients by using a floor-to-floor concept spiced up with cargo unit management that ensured that clients at any point in time knew the exact position and condition of their cargo.

This solution could vastly improve profitability and efficiency of the container line operator and the client logistics under given known market conditions.

However, the container line operator had a very weak management team, who listened more carefully to the fat words of the industry analysts than to the much more precise and less risky recommendations from their Coach/Facilitator.

Although all the employees and business managers were behind our strategic initiative to improve competitiveness, we could not convince the very defensive Strategy Governance Team to continue the container line business.

Looking back, this defensive decision was a very bad one. Another major container line operator has since used a concept similar to the one developed by the business managers and us with great success.

If the Strategy Governance Team comprises people with many different pertinent experiences and skills, there will be many different views on the business opportunities and threats. This multitude of views and experiences allows the team to establish a scope of the Strategic Initiative where threats and opportunities are well balanced in view of the strengths and weaknesses of the enterprise. However, the multitude of views and experiences also presents a risk of negative conflicts among the team members.

Once the Strategy Governance Team has established the initial scope of a Strategic Initiative, it establishes PQA Teams and Workgroup Teams to develop and implement the solution components of the Strategic Initiative. In this respect, the Strategy Governance Team is itself a PQA Team that produces PQA Teams or Workgroup Teams.

The Strategy Governance Team uses a PQA process to establish the underlying PQA Teams with their own scope and management.

The Strategy Governance Team does not interfere with the management of the PQA Teams and the Workgroup Teams once these have been established, if they deliver what is needed and agreed to.

The Strategy Governance Team establishes a Program Office to support the production of requirements specifications and to perform the ongoing support of the Process Governance Teams, the PQA Teams, and the

Workgroup teams working to deliver the required solutions under the Strategic Initiative.

The Strategy Governance Team ensures the "no excuse for failure" conditions of the subordinate teams.

The Strategy Governance Team comprises or is the Change Control Board for the programs and projects under a Strategic Initiative.

The Strategy Governance Team will adapt the Strategic Initiative (govern the strategy) in response to unexpected events and conditions that cannot be coped with on a process and project management level in the PQA and Workgroup Teams supported by the Process Governance Team. Sometimes such changes are required because of changed conditions that were not accounted for in the initial SWOT analysis, for example, changes to corporate ownership, political trends, or corporate management.

Communication is the key to observing and understanding when new Strategic Initiative conditions require major changes to the Strategic Initiative. The communication in the Strategy Governance Team and among the Strategy Governance Team, the Process Governance Teams, the PQA Teams, the Workgroup Teams, the key-stakeholders such as enterprise owners, government, enterprise management, clients and partners, technology representatives, and subject matter experts needs management and methods that will be covered in subsequent chapters.

It is almost too popular to say that all the processes covered in this chapter and the subsequent chapters are communication processes, but it is actually the truth. The organizational constructs and interrelationships allow the communication to take place in an organized and effective way.

2.4.2 Strategy Governance Team Members

The Strategy Governance Team members establish and maintain the detailed scope of the Strategic Initiative from initiation to close out.

Strategy Governance Team members are people who together understand all aspects of the Strategic Initiative. The team members have been thoroughly selected based on their knowledge, skill, and experience and because it is expected that they are motivated for or at least interested in the scope and purpose of the Strategic Initiative.

The members of this team ensure that all known stakeholders are kept happy until final close out of the Strategic Initiative. The knowledge and experience of the team members ensure that together they can define the information and communication that is pertinent to the Strategic Initiative stakeholders of all kinds.

The sponsor has been appointed by the original sponsor based on an often vaguely defined scope and a budget to get the expected benefits realized.

This sponsor manages the Strategy Governance Team, that is, the sponsor handles, directs, motivates, or controls team members to deliver the objectives and to reach envisioned destinations by:

- Ensuring internal efficiency (budget, facilities)
- Doing things right
- Know how
- Managing activities
- Setting tangible targets
- Solving problems
- Having power
- Adapting to events and situations (managing change)
- Organization
- Assigning tasks to members

It is up to the sponsor and the other members of the Strategy Governance Team to ensure visibility and measurability of all accomplished results of the Strategic Initiative in order to ensure continued stakeholder support and budget availability.

2.4.3 The Process Quality Assurance Team

Process Quality Assurance (PQA) Teams are established by Governance Teams to make decisions about what to obtain and deliver, and about how to do it from the start of the Strategic Initiative and in connection with major changes or important milestones.

The PQA Teams lead, manage, and plan activity in order to deliver agreed tangible and measurable solutions to specific requirements.

Such solutions can be factory buildings, complete information systems, a new IT infrastructure organization, a new sales organization, a training program, legal compliance documentation, and many other more or less autonomous solution elements that together contribute to the benefits required from the Strategic Initiative.

The Strategy Governance Team is the top level PQA Team concerned with initiation and kick-off of the Strategic Initiative and is concerned with major changes to the Strategic Initiative as proposed by PQA Teams and Process Governance Teams.

PQA Teams are coached and facilitated by one or more Governance Team members or another competent person appointed by the Sponsor or the Process Governance Team.

The PQA Team can establish one or more subordinated PQA Teams and they appoint managers of Workgroup Teams if required. The managers of Workgroup Teams are most often members of the PQA Team. In the case where the PQA Team appoints a manager of a Workgroup Team, this Workgroup (Project) Manager becomes a member of the PQA Team.

The PQA Teams are the first level of change management. PQA Teams can suggest changes in scope based on new conditions, problems, or risks that they cannot cope with themselves. Such changes are suggested to the Process Governance Team that looks after the PQA Team.

The PQA Team exists only until the full scope under its responsibility has been established and signed off by the people that govern this scope.

Important PQA Team qualities comprise:

- Competence—they can outline the required result quality and design the organization that can accomplish what is needed.
- Knowledge—they cover skills and competence in the different subject matters such as potential market, competition, and technology.
- Experience—they have working experience from the required knowledge areas.
- Team members are complementary to each other, that is, with different backgrounds, attitudes, skills, and competences; if not, you get a biased solution missing important quality elements that might lead to failure of the Strategic Initiative.
- Responsibility—they must take ownership of the Strategic Initiative success factors, organization, and activity.
- Motivation—they drive the Strategic Initiative progress.
- Empathy—they cannot do it alone and they need to communicate in order to get to know and understand the motivating factors of all other important stakeholders.
- Authority—if somebody outside the PQA Team can impose changes not accepted by the PQA Team, then the whole idea of a self-managed team breaks down and the team members lose motivation.
- Good coaching and facilitation.

The Strategic Initiative scope has been defined and signed off by the Strategy Governance Team in terms of:

- Visions
- Objectives
- Quality objects
- Needs/deliverables
- Actions
- Organization
- Risk and other constraints

On this background, the assigned PQA Teams take full responsibility for the delivery of the agreed results, which implies that the PQA Teams are competent and autonomous (Figure 2.6).

The PQA Teams ensure the quality of the delivered solution components with respect to their compliance with the Strategic Initiative success factors.

The PQA Teams establish their own detailed success factors for their delivery of solution components and ensure sign off on their project plan by the Strategy Governance Team or the Process Governance Team looking after their project.

In order to ensure the best possible quality of the defined activities to be handled by the PQA Team, it is recommended that the future Workgroup managers "employed by" the PQA Team are involved as early as possible in the PQA Team, at best in the first PQA Workshop.

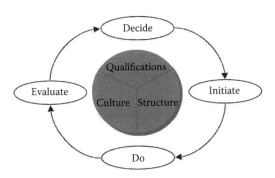

FIGURE 2.6
Team competence.

2.4.4 PQA Team Members

PQA team members are people that together understand all aspects of a solution to be delivered as part of the Strategic Initiative. I recommend the following qualities for PQA team members:

- Three years of middle or senior management position within the line organization or a similar experience and competence that might have been acquired elsewhere.
- Leadership attitude, if possible, that ensures a critical but positive attitude to the business environment and the organization.
- Well respected among peers and higher level managers.
- Empathy and patience as well as skill and experience of Subject Matter Experts (SME).
- Heroes are not welcome if they believe that they have to save the team.
- Original sponsors are welcome, but only in support of the Coach/ Facilitator and his or her sponsor.
- All participants are peers.

Their role is to manage projects and Workgroups that deliver partial solutions to a Strategic Initiative.

Members of PQA Teams handle, direct, motivate, and control resources to deliver objectives and reach agreed destinations by:

- Internal efficiency
- Doing things right
- Know how
- Managing activities
- Setting targets
- Solving problems
- Having power
- Adapting to situations
- Organization
- Assigning tasks

2.4.5 The Process Governance Team

The Process Governance Team is most often set up by one or more PQA Teams and approved by the Strategy Governance Team to be the second change management level in support of the PQA Teams. It ensures on behalf of the Strategy Governance Team that the PQA Teams it supports have "no excuse for failure."

The Process Governance Team performs change management concerned with change requests originating from PQA Teams that cannot themselves cope with new conditions, incidents, observed problems, and risks reported by the Workgroups.

Only the Process Governance Team can request changes from the Strategy Governance Team.

The Process Governance Team exists until all the PQA and Workgroup teams under its governance have delivered their agreed results. The agreed results are what are delivered after all approved changes and adaptations to the original scope have been implemented.

Important Process Governance Team qualities comprise:

- Competence
- Knowledge
- Experience
- Stakeholder trust
- Strong relationship with Strategy Governance Team
- Team members are complementary to each other, that is, they have different backgrounds, attitudes, skills, and competences
- Responsibility
- Motivation
- Empathy
- Authority
- Coaching and facilitation

2.4.6 The Process Governance Team Members

Members in the Process Governance Team are most often also members of the Strategy Governance Team, but besides this they have a deeper interest in the deliverables and the functioning of the PQA Teams and Workgroup Teams they are supporting.

The Process Governance Team members have been thoroughly selected based on their knowledge, skill, and experience and it is expected that they are motivated to coach the PQA Teams and Workgroup Teams they are supporting, especially when it comes to handling of risk and problems.

2.4.7 The Workgroup Team

The Workgroup Teams produce and deliver the solution components required by the PQA Team to which they refer.

The Workgroup Team delivers a tangible result at an agreed cost at an agreed deadline.

Workgroup Teams deliver their results according to precisely agreed Requirements Specifications that allow Accept-Testing and sign off on all results.

The Workgroup Teams are the foundation for the traditional agile principles. The agile demand of physical availability at the same time of critically involved resources for fast development with high quality results can be found in methods such as SCRUM, Xtreme Programming, and others.

It is of course not possible for team members to be together all the time, but it is important to plan the togetherness on specific occasions with important deliverables. Such deliverables are often called Use Stories or Sprints, which simply represent complete business processes with precisely defined input and output quality to be established by the business process.

Strategic Initiative development and implementation normally comprise much other activity and deliverables than software development and software. This implies that rather complex estimation, coordination, and tracking techniques are needed.

Planning and tracking continue to be agile because tracking and progress reporting are based on fully functional deliverables such as:

- Business requirements from Department X
- Workflow description of procedure Y
- COTS system setup of procedure Y
- IT infrastructure setup for security of W
- Intrusion detection ensured for the complete network
- Test models defined for procedure Y ensuring compliance with business requirement X
- Workgroup Z established and producing results as required
- Project plan for PQA Team Y agreed and signed off

The agile principles are pertinent for estimation and tracking. Progress is only measured by delivered solution components that through Accept-Testing have been approved and signed off to work in production.

The Workgroup Team is responsible for the delivery of a result of a certain quality that allows this result to be integrated with the deliverables from other Workgroups.

Workgroup Teams can be dedicated to quality assurance of integrated results from more than one Workgroup Team. Such QA Workgroup Teams have roles dedicated to Accept-Testing, result quality assurance and control, and process evaluation.

Workgroup performance is communicated to all other Workgroups, the PQA Team, the Process Governance Team looking after the PQA Team, and the Program Offices if relevant for tracking and change management.

Project Office project planning and tracking information systems support this communication. However, it is the responsibility of the Workgroup Team manager that this communication is reliable and efficient. The Project Office only ensures integrity and validity of the tracking information from a purely technical point of view, not at all from a project or program management point of view.

2.4.8 Workgroup Team Members

The Workgroup Team Members are skilled and competent resources most often managed by an informal manager or simply managing themselves under the support and coaching of one or more persons appointed by the PQA Team that has established the Workgroup.

The Workgroup Team Members are most often SMEs employed to deliver specific results. The roles always cover development, implementation, and quality assurance concerned with the functional and technical feasibility of their own deliverables.

For Workgroup Team Member roles, I recommend capabilities and qualities such as:

- Documented skill and experience in required subject matters and facilities
- Efficient production of code, solution documentation, software setup, software and solution integration, and solution delivery (to operation and implementation for use)
- Documentation of solution requirements, Accept-Testing, and solution implementation
- Solution and technology operation and support
- Security, compliance, and safety management

The Workgroup Team Members ensure the quality of the products delivered by them concerned with functionality, integrity, and fitness for integration and communication.

2.4.9 Cross Team Membership Value

The same person can be a member of more than one team and the same person can coach/facilitate more than one team.

A team can have more than one Coach/Facilitator in cases where the team objectives are very complex, for example, on the solution side, the technology side, the organization side, or the process side. In this case, SMEs can take the role of Coach/Facilitator. If your Strategic Initiative has a high risk of non-compliance for legal reasons, you might involve a Coach/Facilitator with close contact to legal bodies such as a lobbyist or somebody who is directly a member of the organization responsible for establishing your legal framework.

A Workgroup manager or at least one member of a Workgroup is most often also a member of a PQA Team. Some PQA Team members are also members of the Strategy Governance Team or another PQA Team.

The persons with cross team membership ensure a better interpretation of decisions, success factors, risks and opportunities, and needs across all teams. This ensures that changed conditions and not expected events that will occur all the time are observed faster and more reliably so that feedback to decision making and change management with suggested changes can be effectuated before it is too late to react.

2.5 THE NO EXCUSE FOR FAILURE PRINCIPLE

Once you have succeeded initially to identify the key-stakeholders in your Strategic Initiative, your challenges only begin because such key-stakeholders with the needed competence are difficult to motivate and activate in your Strategic Initiative.

The best key-stakeholders are usually already allocated to or demanded to be involved in other Strategic Initiatives competing with your Strategic Initiative for resources.

The importance of these key-stakeholders in high demand needs to be confirmed in order for them to understand what benefits they can obtain from contributing to your Strategic Initiative.

Facing this stakeholder risk situation, your first response is to ensure that your Strategic Initiative complies with the "no excuse for failure principle":

- You know why you need the key-stakeholder in your Strategic Initiative and you have a whole list of arguments that show the value for this key-stakeholder to contribute to your project.
- You know which internal and external activity that will compete for key-stakeholders with your Strategic Initiative and you respect their importance as well.
- Because you are involving people with very different skills, experiences, and competences, you know that conflicting interests are inevitable. You have organizational elements and procedures in place to avoid conflicts becoming personal with a negative impact on the Strategic Initiative progress.
- By using professional coaching and facilitation, you ensure that conflicts only result in lateral thinking (out of the box) and synergy on workshops and during other types of teamwork.
- At any point in time, teams and key-stakeholders have access to all needed and available resources and knowledge constrained only by accepted limits to their availability.
- You plan to ensure that all required resources to be involved in an activity are available and allocated to the activity before the activity is initiated with assignment of these resources.
- You do not initiate an activity if you know that any required resource is not available to be assigned to the activity.

You incur important risk by not complying with the "no excuse for failure" principle:

- Biased strategy focus because important knowledge or competence that is left out initially might lead to development and implementation of solutions that do not comply with stakeholder needs—you will lose capital and time.
- Key-stakeholders might lose confidence in the Strategic Initiative because the not involved but required resources raise pertinent critiques of the chosen initiative scope and objectives—you will lose time and key-stakeholders might leave the initiative.
- If the involved resources do not have the competence to reach a conclusion about critical success factors and the way forward to an

agreed solution, then the key-stakeholders waste time and lose confidence in (your) management.

- Important processes might be performed with interruptions because of lack of important resources, which leads to waste of time and bad results.
- The initial enthusiasm of the key-stakeholders can disappear very fast if you do not keep them motivated by immediately involving them in pertinent strategic initiative activity, where they get a chance not only to prove their competences, but also to use this competency directly in cooperation with peer stakeholders.
- If the key-stakeholders lose interest in your Strategic Initiative, then the initiative might already have failed.
- If the key-stakeholders get into negative conflict with you or with each other while conducting the Strategic Initiative activity, then the initiative is probably already doomed to fail.
- If some resources accuse other resources of failure, it creates stress and negative conflicts that are the reason for delays that could have been avoided by better selection of resources, better team building, and better working conditions.

Unfortunately, the "no excuse for failure" principle is not all you need to comply with in order to succeed, but if you do not comply with it, you will challenge yourself and the key-stakeholders with unnecessary problems and issues that might in the end make the initiative a fiasco.

The initial dialogue processes and the continued communication with key-stakeholders require serious preparation and very good coaching and facilitation, especially if you are new to such dialogues and communication.

There are more examples of failed Strategic Initiatives than of successful ones—even in major organizations that should have access to the needed and required resources such as Digital, Nokia, Sony, and Philips.

2.6 THE WAR ROOM

The War Room is a construct that I recommend to establish even for teams working geographically separated over very long distances in different time zones (Figure 2.7).

FIGURE 2.7
War room from Americon in Camarillo, California.

The War Room contains all pertinent tools, information, and documentation in support of a specific team with all the technology available to access and manipulate this.

The term "War Room" originates from military combat management. It is the environment where involved officers meet to discuss experience from already executed activity and to make decisions about the activity to perform before the next War Room meeting. In combat, the officers meet very early every morning.

In the case of civil strategic initiatives, the War Room serves the same purpose, but the meetings might happen less often than daily.

I have established War Rooms to be used for Strategic Initiative activity covering decision making, workshops, teamwork, and other events where it is essential that all resources are available in order to ensure the best possible decisions.

I have also established a simple War Room in support of Help Desk activity, where the Help Desk teams met each morning to discuss the support events since the last meeting and to decide on the actions to perform:

- Events to be treated as a project by other organizations
- Events to be resolved immediately by the Help Desk service team
- Open events not yet responded to
- Recurring events demanding future preventive action

Until a team has delivered its result as agreed with the involved stakeholders, the team meets at regular intervals in the War Room to make decisions and to kick off the work processes performed by or governed by the team.

The Workgroup Teams that produce physical solution components can fully benefit from daily War Room meetings. Where I have managed such Workgroups, we have always met in the War Room at 8:30 every day over donuts and coffee to present achieved results and to discuss unexpected or risk managed events and conditions demanding actions to be taken by the team or management.

War Room meetings are never a waste of time, even if you just verify that everything is progressing in good order and with the results as expected, which is of course rarely the case.

When the result or an important milestone has been reached, the War Room is also the place to have champagne and plan communication with other stakeholders.

The War Room is easy to protect physically from unwanted intrusion with cameras and electronic and physical access control. Furthermore, it is easy to build IT firewalls around the work environment of tools, documentation, minutes, and solution components under version control.

There is no physical limit to a War Room. I have had War Rooms as small as the office manager's desktop to a set of interrelated containers with more than 500 m² of space, more than 100 PC workstations, and a combined Workgroup Team with more than 50 people. Only partial teams met every day at 8:30, but they did, and I had no problem getting in contact with my active Workgroup Teams.

2.6.1 The Private Bank Solution Swap War Room (Figure 2.8)

2.6.1.1 Initial Situation

It was impossible to know if the active IT solution could survive more than 12 months, where the 12-month survival could be ensured only by investing in very expensive backup equipment to be thrown away once the new solution became operational.

- The teams in place had to be completely reorganized.
- No competent contractors had been contracted.
- There was no requirements specification because the organization management was convinced that the bought system could provide a solution on its own.

FIGURE 2.8
Container office War Room.

In order to cope with all of these problems at the same time, we decided to integrate all future teams under the same roof while ensuring that all required knowledge and skill was available as needed any time we were in this War Room.

The War Room decision was an agile one, but everyone thought it would be prohibitively expensive because the organization had no "agile" experience.

The negative voices were overruled and the agility was ensured with a comprehensive War Room built with containers backed up with new efficient processes, teams, and documents comprising:

- An initial requirements specification
- New creative contract terms approved by the legal department on concern level
- Contracted needed internal and external skill and competence
- System software and IT operation was established for development, test, and training
- Program and project plans were signed off
- All resources, human and technical, were brought together under the War Room roof
- Security and safety systems and routines were established to protect people and solution components

Only final solution components that had been Accept-Tested to be ready for use were allowed to leave the War Room for business operation.

All work and all intermediate results never left the teams in the War Room.

This team building with War Room resulted in a very successful solution delivery with very motivated team members from inside the bank and from the contractors. Agility was clearly manifested in the work performed (teams of developers, business analysts, and future Information System and IT support and users working closely together) and by the progress tracking that was based on delivered solution components.

Team support for "no excuse for failure" rather than team progress control ensured the fastest possible progress and delivery of fully Accept-Tested solution components.

In the end, the total cost was only half of what a turnkey contractor had offered at the outset.

The alternative contractor offer had been made under the condition that the bank accepted the integrity and full responsibility of the contractor, that is, the contractor decided what the bank needed, which fortunately was not acceptable to the bank director. Just for the record, the bank was willing to pay for this "safe" solution, but it did not want (one more time) to be kept in the dark until a final solution was in place—or failed to be.

2.7 SPONSOR AND KEY-STAKEHOLDER RISK

The Strategic Initiative original sponsor can be a group or a public budget controlled by a selected group of people, often politically selected or appointed by shareholders or other business owners.

Quite often, it is not even obvious who the original sponsor is.

This implies that you and your sponsor without knowing why can be met with fast changing conditions for the Strategic Initiative up to a level where this initiative for some more or less obvious reason is cancelled.

The opposite, for example, changed delivery conditions with more or less time to deliver and improved resources and budget opportunities, is of course also possible.

You are obliged to work "hand in hand" with your sponsor to cope with this original sponsor and core stakeholder risk:

- If you are facing a weak sponsor, you strengthen this sponsor's capability to communicate with your key-stakeholders. This is one of the most important roles of a Coach/Facilitator.
- If you are faced with a strong sponsor, you gain his or her respect and trust in order to make sure that you as coach and facilitator are informed and are asked for advice—even though your sponsor thinks that he or she can act and decide without consulting you.

In my work with major international consultancy organizations and major vendors of software, their top or sales management and sometimes their own smart Coach/Facilitators have quite often tried to establish a direct communication with the original sponsor's core stakeholders (the ones with the budget authority) once these have been identified.

In this way, they bypass the sponsor and his or her Coach/Facilitator. In some cases, they unfortunately enough succeed in getting decisions through that are contradictory to the progress of the Strategic Initiative that they are supposed to support.

The objective of bypassing the sponsor in this way is always to bring in more resources and more product before the Strategic Initiative has come off to a good start or even in the middle of critical development and implementation activity, where the original sponsor is most vulnerable because the first problems, delays, and cost overruns have been reported.

Based on their communication and observations outside the control of your sponsor and you, the original sponsor key-stakeholders might suddenly behave in ways that are detrimental to the foundation of the Strategic Initiative. They express new opinions and attitudes that you do not recognize, but that you need to respond to.

The original sponsors' and other important stakeholders' attitudes and opinions represent risk to the Strategic Initiative that you and your sponsor need to respond to from day one.

One way that you can prevent these quite often catastrophic changes to the Strategic Initiative foundation is by always trying to strengthen your sponsors' ability to communicate in his or her environment of original sponsors, peers, managers, and key-stakeholders.

Person-to-person communication—certainly not by e-mail or telephone—is the best response to sponsor and other stakeholder risk. Such person-to-person communication is not easy to obtain for some important reasons:

- You might not know the person to talk to.
- The person to talk to might not be committed to talk to or listen to you or he or she is too busy to meet with you. (Typical response when you ask for a meeting: "Send me an e-mail with your arguments.")
- The communication is prepared with proofs and arguments that ensure that the original sponsor has trust in you more than in other relationships when it comes to your Strategic Initiative, which is difficult if you do not know the attitude of this original sponsor person.

You cannot win here if you do not have an established relationship with the original sponsor who commits to listening to you and talking with you, a relationship that is established as part of the Strategic Initiative kick off.

The following is an example of a strong (public) sponsor:

While working as Coach/Facilitator in the military facility management program, my sponsor was a very apt officer on the commander level from the navy side.

He knew everyone in high command and he was a vital member of very important non-professional networks.

This sponsor had complete trust in our methods because he had seen them work in another big project, where he was Project Manager.

He trusted the involved resources from my company (still does) and he was able to select the most competent defense resources across army, navy, and air force representing both facility managers and facility users.

He was able to warn us whenever there was a risk of conflict with other interests or projects.

With his intervention, we got access to all the right levels of decision-making and knowledge during the length of the project. Moreover, he ensured that we could present achieved results of more strategic impact on important pertinent conferences and meetings, which allowed us to ensure satisfactory budgets for our required work. Our response was, of course, reliable statistics, delivered results, and happy stakeholders.

The way he appointed people for our workshops and teams on all levels and the way he helped to ensure efficient communication resulted in the success of the teams and in the success of the whole program.

When after 5 years of successful conduct and cooperation our sponsor was replaced with a new one, we got serious trouble because the new sponsor did not at all see my civil company as his future Coach/Facilitator.

Quite often, you will be faced with a de facto sponsor that has limited knowledge and experience from the organizations to be involved with the Strategic Initiative. This is more an opportunity than a threat.

Very experienced and competent sponsors might be less good for the Strategic Initiative than the less experienced but also very competent sponsors might be:

- A sponsor who knows everybody is a bit dangerous if he or she has many prejudices and "friends" that he or she wants to be involved in the Strategic Initiative. If this preferred group does not represent the best available knowledge and experience, you will have a hard time to get the right people "on board." The risk is that your sponsor looses a lot of money before you get him or her on the right track. There is a limit to how you as a coach/facilitator can manipulate your sponsor to do it right the first time because your sponsor does not and will not tell you everything about what he or she knows, and especially not about what he or she knows that he or she does not know.

- A sponsor who initially knows only a few people will be willing to listen to your advice to a much higher degree. He or she will probably use the Strategic Initiative to get to know people. With such an open-minded de facto sponsor, you and the sponsor need to establish initial dialogues with people that you expect will play major roles in the Strategic Initiative. Through these dialogues you gain an opportunity to evaluate the possible participants in the Strategy Governance team with respect to their knowledge, experience, competence, and motivation. This will make it possible to get the best teams in place early and thereby mitigate the risk of delays and loss of money, and it will contribute to a closer relationship between you and your sponsor.

2.8 LESSONS LEARNED

The initial challenge of the strategic initiative sponsor is to formulate the basic need, the initial scope, and to find the competent people that can and will be involved with this strategic initiative to make it successful.

Irrespective of what the strategic initiative is dealing with, which types and sizes of organizations are involved, what kind of people are available, etc. there is only one way to get the initiative off to a good start:

Get the key-stakeholders together and give them no excuse for failure to define the scope of the strategic initiative.

You use communication and remuneration to make stakeholders happy, but in order to make the communication and remuneration successful you need to know the stakeholders and their needs and expectations.

The sponsor initiates the strategic initiative and signs off on the scope as this is established originally and as it is adapted to new conditions and events during the lifecycle of the strategic initiative.

The Coach/Facilitator delivers the quality system and supports all directly involved team members and managers with appropriate guidelines, procedures, and documentation standards during the length of the strategic initiative.

The unknown unknowns are the conditions that you might discover as group synergy or by simple luck, but you will not find them if the minds of you and your stakeholders have not been set to be observant and creative—to "think out of the box."

Although the preconditions are important and we have to know and understand them, they are historical. Requirements specifications, wish lists, and even contracts are merely guidelines to get the initiative started off.

This initial dialogue with potential key-stakeholders who are potential participants in the Strategy Governance team makes it possible for us to answer the why question with focus on real pertinent opportunities and threats.

The Project Offices support all teams that run projects in the corporate portfolio of projects, as well as the projects that belong to Strategic Initiatives or programs.

The Program Office is a secretarial facility and knowledge management center for a specific program or Strategic Initiative.

The best solution architectures, requirements, and designs emerge from self-organizing teams. Self-organizing does not ensure this situation; it is the fact that the teams are established in such a way that self-organizing is possible to ensure the situation.

Once the Strategy Governance team has established the initial scope of a Strategic Initiative, it establishes PQA Teams and Workgroup Teams to develop and implement the solution components of the strategic initiative.

The Strategy Governance team establishes a Program Office to support the production of requirements specifications and to perform the ongoing support of the Process Governance teams, the PQA teams, and the Workgroup teams working to deliver the required solutions under the Strategic Initiative.

Communication is the key to observe and understand when new Strategic Initiative conditions require major changes to the Strategic Initiative.

PQA teams make decisions about what to obtain and deliver, and about how to do it from the start of the strategic initiative and in connection with major changes or important milestones.

The Process Governance team is most often set up by one or more PQA teams and approved by the Strategy Governance team to be the second change management level in support of the PQA teams.

The Workgroup teams produce and deliver the solution components required by the PQA team to which they refer.

The initial dialogue processes and the continued communication with key-stakeholders require serious preparation and very good coaching and facilitation.

The War Room contains all pertinent tools, information, and documentation in support of a specific team with all the technology available to access and manipulate this.

The original sponsors' and other important stakeholders' attitudes and opinions represent risk to the strategic initiative that you and your sponsor need to respond to from day one.

3

Strategy Process Quality Management

Once our Strategic Initiative teams have been established, it is time for them to get to work, that is, to plan the delivery of the strategically aligned solution and to deliver it by executing the plans.

In order to succeed with the solution delivery, the teams establish a complete set of plans that with the highest possible probability lead to solutions accepted by the stakeholders.

Planning and plan execution of Strategic Initiatives is not just Project Management; it is, to an even higher degree, Risk Management:

> The objective of Strategic Initiatives is to reach FUTURE situations and conditions with high PROBALITY that will provide the IMPACT wanted by the involved stakeholders.

Strategic Initiatives are risk. They can fail or succeed. In order to optimize the chance or probability of success, we apply risk management to the Strategic Initiatives. Project Management on its own will not do the job.

Risk Management performed efficiently can allow the teams involved with Strategic Initiatives to build plans that have a higher probability to achieve the solutions and results (the impact) demanded by the stakeholders.

In this respect, Strategic Initiative Process Quality Assurance (PQA) is Risk Management and in the work that we have performed, Risk Management is PQA.

PQA and Risk Management adapt to the same basic definition. Risk Management comprises methods, standards, tools, and techniques that allow the performing teams to:

- Identify opportunities (risk with positive impact on stakeholder benefits) and threats (risk with negative impacts on stakeholder benefits).

- Analyze the risks for their cause and impact.
- Evaluate and quantify the risks with respect to their probability and their impact.
- Respond to opportunities by planning for them to come through.
- Respond to threats by planning for them to be removed or to be mitigated.

Risk Management processes are not finite; they continue throughout the Strategic Initiative and help the teams to survey the risk situation of the initiative.

Please remember that we are always faced with pertinent unknown unknowns and unknown knowns that are ready to surface at any point in time in the future of our Strategic Initiative.

Risk responses are always built into the project plans. You respond to risk by adapting your plans to:

- Accommodate the best possible resources.
- Utilize the best possible procedures, standards, and techniques.
- Adapt the solution to be SMART (Specific, Measurable, Achievable, Realistic, Time bound).
- Ensure satisfactory funding by efficient stakeholder communication.

I admit that I am not in favor of artificial elements such as Residual Risk and Contingency Reserves based on more or less good exposure calculation:

Exposure = Σ(Impact – value $\times$ Probability of Risk – event or condition)

Personally, I prefer to call things by their real name:

- Simply opportunities and threats—they are there if they are documented in the risk list.
- Funding instead of reserves—funding is decided by the sponsor and will depend on factors other than risk, such as trust, need, and capability. I have never met a sponsor who provided "free" reserves to any project manager, but funding after serious negotiation, yes!

The symbiosis of PQA and Project Management is well established in many corporate standards, but funny enough not in the PMI and Prince standards, where Risk Management is defined as a sub-management process to Project Management.

One of the major Swedish international concerns in the Information and Communication Technology industry uses Risk Management as their initial process and of course continues to apply Risk Management to Planning and Change Management in the full lifecycle of their projects and programs.

I had the opportunity to train their development engineers in PQA. It was in this context that we discussed how Risk Management should be applied to Project Management.

The Strategy Governance Team establishes and maintains the master plan that binds together and integrates all the subsequent plans. The Strategy Governance Team establishes the master plan as the top level PQA Team. In this top level PQA process, they establish the first level of subordinate PQA Teams that together ensure the delivery of the required strategically aligned solution components.

PQA is used to ensure the quality of the initial plans, but it is also used to ensure the quality of changed plans, especially in connection with PQA review workshops that are used to respond to risk and to adapt the plans to required changes decided by the Strategy Governance team.

Conceptually PQA always starts with the opportunities. The logic of this is that if you cannot find a feasible plan based on opportunities, there is no need to look for risk because the Strategic Initiative is already doomed to fail. If, on the other hand, you have at least one way to ensure the success of the Strategic Initiative, you have a sound base for optimization by searching for and adapting to other opportunities and risks. PQA works like this:

- In order to establish a feasible plan, PQA initially looks primarily at opportunities to reach the targets. This opportunity-based plan has a very high probability to deliver results that are what the stakeholders want; that is, if the threats were not there, which of course they are.
- Under the planning of this opportunity-based project or Strategic Initiative, the stakeholders and resources to be involved with initiative execution uncover the threats and sometimes new opportunities, but the focus here is threat identification and threat analysis.
- The opportunity-based plan is adapted to respond to the threats in order to ensure the highest possible probability of success from the Strategic Initiative, which is continually adapted to the future risk situation.

PQA is the method for Strategic Initiative establishment and planning based on intensive teamwork in the PQA Teams with brainstorming that documents the agreed Strategic Initiative for implementation.

Agreement to the established objectives for the Strategic Initiative is ensured by active involvement of all pertinent and available knowledge from inside and outside of the organization in the PQA process.

PQA ensures:

- Identification of the Strategy sponsors and other key-stakeholders
- Establishment of the agreed strategy with detailed objectives, Strategic Initiatives, teams and team organization, management, and communication that can ensure the success of the strategy under fast changing conditions and high risk
- Establishment of standards for processes and documentation that can answer these basic questions:
  - Where we are?
  - Where do we want to go?
  - Why do we want to go there?
  - How do we want to go there?
- The answers are given in terms that can be easily interpreted and agreed to by all involved stakeholders
- Implementation of the strategy with timely execution of decided Strategic Initiatives, timely measurement and approval of results and benefits, and efficient change management in support of strategy governance

3.1 STRATEGY QUALITY MANAGEMENT

The PQA method comprises a proven set of tools and techniques that are used to assure the quality of Strategic Initiative processes. However, PQA cannot ensure the quality of the entire strategy on its own. For example, a simple question such as "What is the status of the Strategic Initiative progress?" cannot be answered by PQA. The reason for this is that PQA as a Risk Management-based method is looking ahead. PQA can tell you what the probability is to be on the right way to success in two weeks if this or that opportunity is pursued and this and that risk is avoided or responded to well.

You need supplementary procedures and tools to PQA to perform Strategy Quality Management (SQM), for example, if you want to know where you are compared to where you want to be while executing a Strategic Initiative:

- Project and program tracking is based on a number of performance indicators that can tell you if an activity is delayed, if the project or program is delayed, and quite often if this delay is curable, that is, if changes to the baseline are needed.
- Analysis of information system requirements and solution design is a strong foundation for time and cost estimation. The developers and implementers of the solution can give you reliable feedback about solution quality progress related to usage of resources and funds that can be used for change management, where PQA comes back into the picture.
- Breaking the delivery down into manageable Work Packages based on easily verifiable use cases that are not started, in production, or delivered makes it possible to build Work Breakdown structures for agile solution development and implementation that ensure visible progress and fast and reliable reactions to change requests.

Other performance indicators can be based on:

- Deadlines for access to needed resources and sufficient infrastructure
- The quality of solutions based on, for example, number of failed tests and outstanding errors
- Number of outstanding compliance rules to adapt to
- Contractor behavior versus Scope of Work
- Earned value versus planned costs

The methods, tools, and techniques that we use to ensure the quality of the strategy over its entire lifecycle besides PQA comprise the following:

- SWOT (Strength, Weakness, Opportunity, and Threat) analysis is performed to fully understand the conditions of the corporation (market, organization, finance, technology, etc.). The SWOT analysis, if well performed, allows corporate leadership (the strategy sponsor) to establish the initial scope of the strategy. The initial scope comprises:
  - The corporate situation in its market—both current and potential
  - Competition—both current and potential

- - The strategy stakeholders
  - The strategy objectives
- Why is always explained.
- SWOT is used in the preparation of PQA. You need to know where you are in order to be able to navigate to where you want to be.
- In the initial phase of Strategic Initiatives, the scope of the strategy does not need to be very precisely formulated because one of the objectives of PQA is to define the precise scope of the strategy under future conditions that are not known while the initial Strategic Initiatives are established.
- However, it is important to understand the situation of the strategy (the why) in order to be able to choose the most pertinent stakeholders for the "no excuse for failure" strategy, and in order to be able to verify later on that the reason for the strategy is still relevant. If the strategy is based on fear for competition that has ceased to exist, it might be time to build a more aggressive strategy.
- Strategy implementation and governance use classic portfolio, project, and program management methods to plan, schedule, and track the progress of solution implementation. In this respect, we are compliant with PMI and Prince Standards and to the best of my knowledge with the Swiss Hermes Standard.
- In this context, procurement and contracting of external resources for implementation of COTS-based solutions is covered by an example without any discussion of general legal terms and conditions such as it is done in, for example, the PMI PMBOK standard. The sole purpose here is to ensure good team building and low risk delivery of results where sub-contractors are involved.
- Change Management and PQA used in the context of portfolio, project, and program management ensure timely adaptation of the strategy as required in light of unexpected results, conditions, and events.
- Under delivery of strategy solution components, the Coffee Bean methods are used. The Coffee Bean methods ensure ongoing and timely control with the development and implementation of tangible deliverables agreed to by the strategy sponsor and the teams involved with the Strategic Initiatives. The Coffee Bean offers a consistent framework for the Project and Program Work Breakdown Structures. It comprises the standards and methods for Business

Analysis (Information Requirements Study), Solution Design (Object Lifecycle Analysis), Agile Solution Development, and Solution Implementation (Accept-Testing).

3.2 STRATEGY QUALITY OBJECTS

It is obvious that the needs of corporate stakeholders are very different, and most often inconsistent and contradictory. Nonetheless, corporate leadership and management establish a strategy that they can defend, that is, that can be explained to and accepted by any stakeholder.

In order to solve this task, a set of commonly accepted strategy quality objects are required for strategy establishment. These quality objects have the following characteristics:

- They are tangible.
- They are measurable.
- Their validity and pertinence is agreed to by all stakeholders.
- The quality objects can express all stakeholder needs.

The three basic quality object classes with which we work are:

- Solution
- Process
- Organization

The three quality object classes represent a complete classification of elements that are explicitly defined for any project, program, or business activity performing partial or complete implementation of the corporate strategy.

- **The solution** can be any combination of products, infrastructure, technology, situations, conditions, business processes, requirements specifications, and organizational elements required to satisfy the needs of the stakeholders.
- **The process** comprises all activity required to develop and implement (deliver) the solution in time and in compliance with pertinent internal and external standards and legal conditions. The process allows the corporation to optimize the tools, methods, and procedures used in order to obtain the best possible efficiency, for example, using agile and Lean principles whenever possible.

- **The organization** encompasses anyone directly involved with establishing and implementing the corporate strategy whether internal employees and managers, external consultants, subject matter experts, sub-contractors, or legal and public bodies such as local and national government and unions. The structure of the organization ensures the opportunity to get a beneficial classification of business or working processes and their demand for skills and competences, while the culture of the organization caters to the extremely important communication required to lead, manage, plan, and execute Strategic Initiatives to deliver expected results and benefits.

PQA and other SQM processes such as outlined above ensure completeness and sufficiency of the solution, the process, and the organization required to achieve the benefits expected from implementation of the strategy.

In the context of strategy quality, we use Success Factors and Critical Success Factors (CSF) to establish a valid and complete definition of all quality objects' attributes.

A success factor expresses a pertinent wanted and needed attribute of one or more quality object, while a CSF is a class of success factors. Success factors cannot be processes or questions; they describe an attractive situation or event, which forms part of the stakeholder's visions.

3.2.1 Strategy Quality Object Attributes

The three basic quality object classes can be further subdivided into more specific quality object attribute types that are addressed by the corporate strategy and that make it easier to understand and evaluate the completeness of the corporate strategy for the involved stakeholders.

The attribute types can be specific to the strategy situation, but in most cases, they will fall under the nine attribute types shown in Figure 3.1.

During the PQA-based definition of success factors and CSF, we use a simple rule to control the quality of the success factors:

If any quality object or pertinent attribute is left undefined by a strategic requirement, the resulting strategy will probably not be sufficient.

The PQA team cannot ensure that the strategy as implemented will lead to the expected result if the success factors found under PQA are insufficient or missing. The biggest risk here is that the PQA team does not

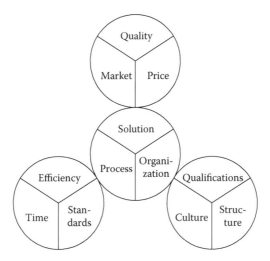

FIGURE 3.1
Strategy quality object classes and attributes.

discover the insufficiencies or the missing success factors before it is too late. In order to prevent or mitigate this risk, it is highly recommended to use an independent facilitator to facilitate the PQA process and to ensure its quality.

If it is explicitly stated that an undefined quality object or attribute is supposed to be handled as "within known opportunities," then all stakeholders understand why nothing further is done or needed concerning this quality object.

A high-quality corporate strategy defines all agreed quality object attributes (here, nine in total) precisely with the help of the agreed success factors and the CSF resulting from PQA performed by the PQA Teams involved with the Strategic Initiatives.

3.3 SQM FACILITATION AND INITIATION

SQM is initiated by establishment of the corporate strategic framework that consists of:

- Strategy Sponsor
- Key-stakeholders
- Strategy Governance Team

- The corporate situation of SWOT
- Corporate objectives
- Corporate quality standards
- Facilitation

It is a common mistake to confound corporate strategy with all ongoing and planned activity whether managed in the form of projects and programs or by the line organization management as ongoing business activity.

It is explicitly declared by the corporate Strategy Governance Team or a similar body with the same strategic responsibility that the ongoing planned activities form part of the corporate strategy. In this context, the ongoing planned activities have agreed explicit objectives for their contribution to the corporate strategy success factors.

Day to day business activity governed popularly speaking "bottom up" by line or functional management cannot replace a strategy defined by leadership top down. Leadership will establish strategic objectives that will inspire management to optimize their business operations, but that does not mean that the sum of the business operations as performed equalize the strategy.

The Strategy Governance Team comprises as a minimum the strategy sponsor coached by an experienced facilitator.

At best, the Strategy Governance Team represents corporate leadership and consists in this respect of a team of managers with the power and authority to initiate, implement, evaluate, and sign off the Strategic Initiatives to be established.

The Strategy Governance Team knows and understands the conditions and the needs of the corporation in its current situation based on a thorough SWOT analysis or based on the mutual knowledge and experience of the team members.

An experienced facilitator coaches a SWOT analysis in order to ensure a feasible result. The facilitator is at best an independent highly qualified person who is fully trusted by the Strategy Governance team.

The facilitator has at least the following qualifications:

- Is not an employee of the corporation
- Knows PQA and other SQM methods in depth based on own practice
- Has good basic industry knowledge, but is not an expert (or a hero)
- Has strong communication skills
- Has strong empathy
- Has natural authority

The Strategy Governance Team defines the initial strategy scope and objectives based on the individual visions and expectations of the team members.

The strategy is initiated and later on changed if required by using PQA. In this respect, the Strategy Governance Team takes on the role of the top level PQA team.

3.4 THE PQA METHOD

PQA is the set of processes that are used for establishment of the Strategic Initiatives of the strategy:

- The introduction to the PQA process for the PQA team that performs the PQA processes under a Strategic Initiative
- The PQA Team PQA Workshop that is a brainstorming-based team-building process that results in:
  - Individual visions and missions documented and discussed in support of conflict prevention and establishment of mutual respect
  - Agreed Success Factors defined and approved to be true and complete
  - Critical Success Factors defined and mutually agreed to not leaving one single success factor non-classified
  - Required activities to be performed
  - Appointment of PQA Team members to lead the required activities
  - A plan for PQA result distribution and future reviews of the planning effort concerned with the agreed activities under the Strategic Initiative
- Production and distribution of the PQA Workshop result to the strategy sponsor and the PQA Team members
- Production by the appointed activity leader of the detailed Activity Description for the activity identified in the workshop. These Activity Descriptions are reviewed for compliance with the Activity Description standard by the Project Office and distributed to the PQA Team for review as agreed to in the PQA Workshop
- In a series of review meetings or in a dedicated strategic planning workshop, the suggested activities are approved for initiation and execution. During these reviews, Risk Management is used to

identify, evaluate, and define responses to identified risk. The risk responses are improved ways to perform already defined activities combined with identification of new required activities to mitigate threats and pursue opportunities.

The activities are planned in detail once approved. Duration and resource needs and possible costs and dates of required result delivery are estimated with due attention to already executing activity, activity interdependencies, and availability of required resources.

The resulting project plans are approved by the Process Governance Team, which performs the role of process or project sponsor. On strategy top-level, the sponsor is the Strategy Sponsor or the Strategy Governance team, while low-level projects performed by Workgroups can have sponsors appointed by higher level PQA Teams such as the Strategy Governance Team.

Major programs such as the implementation of an atomic power plant, the implementation of a ship's wharf, or the establishment of a new bank operation will have sponsors and Process Governance Teams with huge autonomous budgets on many levels.

Once executing, the Strategic Initiatives are tracked for progress and quality of delivered results. The status of the Strategic Initiatives is communicated according to the communication plan, not only periodically such as ongoing business operation reporting, but also such as when pertinent deviations from expected conditions are observed.

Important deviations from initial expectation as evaluated by the Process and the Strategy Governance Teams can lead to major changes of the strategy that require new PQA processes and new Strategic Initiatives.

3.4.1 PQA for Destructive Conflict Prevention and Stakeholder Motivation

One of the most important values of PQA is that it can prevent destructive conflicts and improve stakeholder motivation. All stakeholders are motivated for their own personal reasons. PQA visualizes the motivating factors of the key-stakeholders by committing them to document and present their personal vision of the result of and their personal view on their mission during the Strategic Initiative. Furthermore, the key-stakeholder motivation is kept alive and even strengthened by involving them actively in the implementation of successful solutions.

PQA preparation and execution require strategic thinking, planning, and communication in order to establish the commonly accepted (initial) objectives.

The common acceptance of initial objectives and the initial strategy of activities to achieve these objectives ensures an opportunity for a "no conflict" implementation process. The "no conflict" implementation process can be realized long term if it comprises efficient change management.

Efficient change management implies:

- Processes that allow fast and precise tracking (deviation reporting) of all progress
- Procedures that allow deviations to be correctly interpreted and reported
- Procedures that ensure fast and correct response to problems, new threats, and new opportunities

When building a bridge, common targets are obvious, which means that the "no destructive conflict" solution is more process- and organization-based, which in most cases can be controlled by management.

When building IT and Information System solutions, common targets are much less obvious. Quite often, the resulting solution that can be accepted by the stakeholders from such projects does not reveal itself until the final Accept-Testing.

It therefore becomes a key to the "no destructive conflict" situation that IT and Information System solution targets are thoroughly defined, governed (adapted), and agreed to by all stakeholders. This is especially important when we talk about "moving targets." Agility is necessary.

Besides well-defined agreed targets, the required processes to reach the targets are thoroughly defined:

- Project Requirements Document
- Risk Management
- Tracking standard key figures
- Change management organization
- Change management procedure
- Change implementation procedure

The organization that can execute the processes and deliver the required results is established and governed in such a way that the "no destructive conflict" situation can and will be reached and communicated throughout the implementation of the strategy.

PQA ensures a strong team feeling among the PQA Team members, which strengthens their ability to make decisions, and to cooperate and negotiate among themselves and with external partners.

3.4.2 The PQA Teams and the Cascading PQA Processes

The Strategy Governance Team is by definition the first PQA team. The Strategy Governance Team defines:

- The initial success factors and CSF of the Strategic Initiative
- The main activities required
- The key-stakeholder who is responsible for the success of the activity (the activity leader or the coach of the future activity leader)

The activities defined on Strategy Governance Team level are most often in themselves so comprehensive that they demand their own PQA Team to perform a more detailed PQA process for their establishment with more detailed success factors, deliverable requirements, activities, management, and team members.

The lowest level PQA Teams define activities to be performed by Workgroups that produce tangible deliverables in support of the finally agreed upon strategy benefits.

Based on a cascading PQA process, the Strategic Initiative scope will be fully defined with:

- Expected benefits that can be measured
- Solution components that are agreed to
- Tangible deliverables that allow performance measurement
- Teams and team members and their roles
- Organization for communication and change management

During the PQA-based development of the Strategic Initiatives, all PQA Team members are peers irrespective of their origin within or outside the corporation in question, that is, they have been chosen because they can:

- Formulate the real need of the organization and the benefits to be obtained (WHY)
- Describe the different solution opportunities and their consequences (WHAT)

- Describe the processes and the required resources for a successful result (HOW)
- Select a solution and assign the required resources to achieve it (WHEN)

3.5 PREPARE THE PQA WORKSHOP WAR ROOM

PQA Workshops and Workshop reviews need space for the number of participating people to be seated around one oval table that allows all participants to see and hear each other.

Let us call the meeting room for the PQA Workshop and the PQA Reviews a War Room.

The PQA War Room for initial PQA Workshops requires the following equipment:

- One oval (super-ellipse) or round table
- Flipchart sheets
- Big white board
- PowerPoint presentation equipment
- Walls or windows where you can glue the filled-in flipchart sheets
- Writing equipment for paper and flipchart sheets
- Notepaper for all participants
- Coffee and donut service all the time just outside the War Room
- Lunch service just outside the War Room

In addition to these war room facilities, for PQA reviews you need access to the following:

- Communication facilities that allow you to call and talk to team members and people to be involved in activities and who cannot attend the review meeting. You will need a virtual meeting environment such as Skype, but with good white board facilities.
- The corporate project offices information systems for planning and scheduling.

The War Room can be inside the corporate premises, which is fine for an intensive one-day workshop and disciplined participants.

- No one leaves the workshop outside the agreed coffee and lunch breaks.
- You continue until all activities have been lined up.

If the Strategic Initiative has enough importance, priority, and complexity, you should allow two days for the initial PQA Workshop and do it in a hotel or another professional meeting environment. The rules of workshop attendance are the same.

3.5.1 Running a One-Day PQA Workshop

A one-day workshop can have the following agenda:

8:30–8:45	Welcome coffee
8:45–9:15	Introduction to participants with Question and Answer (Q&A) session
9:15–9:30	Coffee break
9:30–10:30	Vision and mission presentation
10:30–10:45	Coffee break
10:45–12:30	Success Factor suggestion
12:30–1:30	Lunch; the facilitator evaluates the suggested Success Factors
1:30–2:30	Success Factor suggestions continued if needed until no more suggestions
2:30–2:45	Coffee with cake
2:45–4:30	Definition of Critical Success Factors
4:30–4:45	Coffee break
4:45–5:45	Outline required activities and evaluate completeness until completeness has been achieved with a good level of the activities
5:45–	Close out with review meeting planning and assignment of responsibility for Activity Description

It is you as facilitator who decides when the right quality of the Success Factors, the Critical Success Factors, the decided activities, the activity responsibility, and the review planning has been reached.

However, it is the PQA Team members who formulate and agree on all elements; as facilitator, you only coach and facilitate.

3.5.2 Running a Two-Day PQA Workshop

The agenda is almost the same as the one-day workshop, but in addition, you plan for an evening activity. If you want a high-quality Day 2, you should keep the alcohol consumption to a minimum during the evening activity, which is possible with physical exercises such as dancing lessons or friendly competition bringing the team members even closer together.

A two-day workshop can have the following agenda:

DAY 1:

8:30–8:45	Welcome coffee
8:45–9:15	Introduction to participants with Q&A session
9:15–9:30	Coffee break
9:30–11:00	Vision and mission presentation
11:00–11:15	Coffee break
11:15–12:30	Success Factor suggestion
12:30–1:30	Lunch
1:30–2:30	Success Factor suggestion continued
2:30–2:45	Coffee with cake
2:45–4:30	Success Factor suggestion continued
4:30–4:45	Coffee break
4:45–5:45	Success Factor suggestion continued until no more suggestions
5:45–7:00	Facilitator evaluates the suggested success factors and prepares flipchart sheet for Critical Success Factor definition
7:00–10:00	Dinner followed by whatever you want—games, dancing, wine tasting, etc.

DAY 2:

8:45–9:15	Presentation of the results of Day 1 and eventually adding of new suggested Success Factors with evaluation
9:15–9:30	Coffee break
9:30–11:00	Definition of Critical Success Factors
11:00–11:15	Coffee break
11:15–12:30	Definition of Critical Success Factors
12:30–1:30	Lunch
1:30–2:30	Outline required activities and evaluate completeness
2:30–2:45	Coffee with cake
2:45–4:30	Outline required activities and evaluate completeness until completeness has been achieved with a good level of the activities
4:30–4:45	Coffee break
4:45–5:45	Continue outline of required activities and evaluate completeness until completeness has been achieved with a good level of the activities

5:45–   Close out with review meeting planning and assignment of responsibility for Activity Description

Again, it is you as facilitator who decides when the right quality of the Success Factors, the Critical Success Factors, the decided activities, the activity responsibility, and the review planning has been reached.

3.6 PREPARE THE PQA TEAM FOR PQA AND PQA WORKSHOP

The PQA Workshop is a brainstorming session that follows a specific set of rules of conduct. All invited participants have been selected because of their assumed interest in the subject of the Workshop, and because of their competences in the subject matter.

The PQA rules have been established in order to ensure active involvement of all participants. No one is accepted just as a guest or to listen in passively. On the contrary, the rules are there to ensure the synergy that is only possible if the explicit and tacit knowledge of all participants are provoked to be used.

In order to set the mind of the participants in the PQA Workshop, give them detailed information about:

- The current situation of the involved organizations
- The known objectives of the Strategic Initiative to be planned
- The other team members
- Why and how the brainstorming is performed

This information is presented in two ways:

1. A written introduction prepared by the facilitator and the sponsor and mailed to all participants in the PQA Workshop one to two weeks before the workshop
2. A PowerPoint presentation that is used for opening the workshop by presenting:
   Why the workshop is conducted
   Why the participants have been selected
   Why the workshop rules of conduct are the way they are

The PowerPoint presentation illustrates what has already been presented in the written introduction and allows the facilitator and the sponsors to present themselves and to explain why the PQA Team members have been chosen.

You also have to take into account that the PQA Team members might have had so little time to prepare themselves for the workshop that they have not read the written introduction.

It is very important for the quality of the workshop result that all participants get a chance to set their mind on their workshop involvement before the start of the brainstorming elements.

The PowerPoint supported presentation is also used to answer questions from the participants before the workshop starts. Most often participants have already tried another brainstorming method, and they would like to know why your method is different.

Some participants might have bad experiences from other workshops and some might have had good experiences.

In these cases, I explain that there has been no bad experiences with the PQA Workshop because:

- The workshop is used directly for decision making, not for recommendations that might afterward be ignored or rejected by management.
- PQA is fully backed up by and signed off by management.
- Mutual respect is established in order to prevent destructive conflicts and to allow more than one point of view or personal opinion about a subject.
- Discussion among participants is promoted in order to ensure the best possible value and formulation of Success Factors and Critical Success Factors.
- The selection of participants has been done by the sponsors after thorough evaluation of the pertinence of their skills and competences.
- The workshop initial scope and objectives have been thoroughly documented before the workshop.
- There are strict rules of conduct that allow synergy and that ensure that all participants contribute in an optimal way.

You prepare the written introduction to PQA and the PQA workshop as Coach/Facilitator together with your Sponsor.

If there are key-stakeholders on organizational levels above your Sponsor, it is highly recommended to let these stakeholders sign off on the

PQA Workshop introduction in order to avoid any objections later that will have a very negative effect on the PQA Team member motivation.

One of the keys to successful conduct of a Strategic Initiative is that the top-level stakeholders and the original sponsors are involved and activated in the initiative activity whenever this gives them an opportunity to show their personal motivation and their support of the PQA Teams. At best, this involvement gives top-level stakeholders and sponsors an opportunity to proactively promote the importance of their Strategic Initiative.

In most PQA Workshops, these high-level stakeholders (most often a general manager or another member of the board of directors that is also the original sponsor) open the PQA workshop with a short personal speech that explains the importance of the workshop result to the participants.

After the workshop, the original sponsors can evaluate the PQA Workshop result.

On one specific PQA Workshop of high strategic importance where we performed the initial PQA for the pharmaceutical factory implementation, the future board of directors' team of the factory visited the 2-day workshop after the first day and gave the PQA Team very positive feedback on the intermediate result with visions and Success Factors.

This intermediate feedback from key-stakeholders and original sponsors halfway through the PQA Workshop contributed to the motivation of the participants and the quality of the successful result of the PQA Workshop.

3.7 THE PQA PROCESS AND WORKSHOP INTRODUCTION

The PQA process and Workshop introduction comprise two parts:

- A specific part that describes the condition of a specific workshop:
  - Where and when
  - Who
  - Scope and objective (Why, What, Constraints)
  - The roles and the responsibilities of the participants
  - How the participants can prepare themselves for PQA involvement
- A common part that explains the rules of conduct of the PQA Workshop and the activity performed after the Workshop:

- How the introduction to PQA and PQA Workshop has been prepared
- What happens in the PQA Workshop
- How the PQA Workshop result is produced and presented
- What an Activity Description is and why it is produced after the Workshop
- PQA Team activities performed after the Workshop

3.7.1 PQA and PQA Workshop Introduction Examples

The PQA and PQA Workshop Introduction examples introduce you to what a simple introduction can look like and the common part that is the same for all PQA processes unless very special conditions demand another way to perform PQA, such as you will see with the private bank information system swap case.

An example of a standard PQA Introduction is shown in Appendix A. A Powerpoint presentation can be downloaded from www. lyngso.lu.

3.8 THE PQA WORKSHOP

The PQA Workshop has the following agenda:

1. Welcome and PQA introduction to the PQA Team
2. Question and Answer session (short)
3. Participants define their personal visions and mission statements
4. Definition of Success Factors and Success Factor quality control
5. Classification of Success Factors into Critical Success Factors
6. Setup of the PQA Matrix
7. Outline definition of activities and activity quality control
8. Presentation of the Activity Description
9. Assignment of activity leaders to all outlined activities
10. Planning of PQA review activity

Each element is described in the written PQA and PQA Workshop introduction and in the PowerPoint-based PQA Team PQA introduction.

Here I will take the opportunity to explain in more depth why we have chosen to conduct PQA the way we have.

3.8.1 The PowerPoint-Based PQA Team PQA Introduction

The PowerPoint introduction is used to make sure that all participants understand why the PQA Workshop is performed and why they have been invited as members of a PQA Team to perform PQA.

The verbal introduction can be performed by the sponsor or the facilitator or by both. If possible, the original sponsor or another top-level key-stakeholder should open the workshop to show how important it is to arrive at a good result.

The first slide presents the participants (Figure 3.2).

As facilitator, you have asked the original sponsor to open the Workshop with a statement about how important the expected PQA result is for the enterprise.

For the original sponsor, the PQA result is a first step to fulfill the expectations and objectives of the enterprise, the establishment of the full Strategic Initiative scope. The sponsor knows what the PQA result looks like because you have informed him or her about this and because the sponsor has had the opportunity to read the written PQA Introduction.

The result seen from the PQA Team point of view is a project plan that outlines the solution components, defines the processes and activities required to establish the solution components, and establishes the organization to perform these activities; however, the team might not be fully aware of this when you start the workshop.

Your objective is to make sure that all participants understand why and how the PQA Workshop is conducted. The participants are invited to ask

Strategic Initiative PQA Workshop

- Participants
  - Joe, Production Manager
  - Peter, HR Manager
  - Anita, Sales Manager
  - Paul, Support Manager
  - Erica, Business Development Manager
  - Lisa, IT Manager

- Facilitator
  - Andrew

FIGURE 3.2
PQA participants.

questions about the written introduction and whenever they want a PQA element explained more thoroughly.

You explain why the PQA participants have been selected supported by a slide that shows how we intend to use their explicit and tacit knowledge to establish the best possible plan for establishment of the Strategic Initiative (Figure 3.3).

It is your opportunity to tell the participants that we want their knowledge to be visualized and shared in the context of the Strategic Initiative. You explain what your role as facilitator is and what their roles as planners and decision makers are.

Before you ask the participants to define their vision and mission with respect to the Strategic Initiative, you can introduce what you mean by Success Factors and Critical Success Factors. The participants will use the success factors to document the initial requirements to Solution, Process, and Organization in order for these objects to deliver the expected result of the Strategic Initiative (Figure 3.4).

Some participants might have an idea about what success factors are, but you need to explain how the success factors are used under PQA.

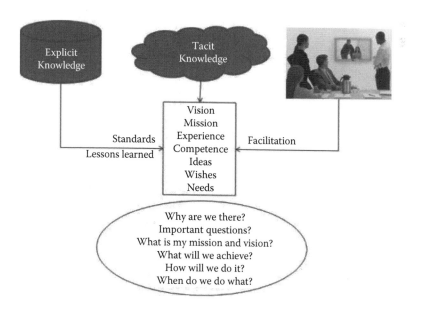

FIGURE 3.3
PQA knowledge sharing.

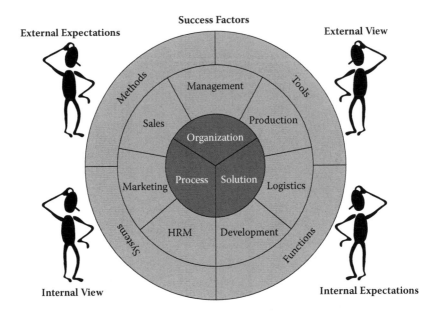

FIGURE 3.4
Success Factor elements.

3.8.2 Question and Answer Session

In some cases, during the introduction presentation or later on during the PQA Workshop process you will be met by the following question:

"When talking about organization, do you mean the organization of the Strategic Initiative or do you mean the corporate organization?"

The answer is not straightforward, but at the same time it is quite clear:

- The Strategic Initiative organization is a part of the corporate organization, so Success Factors addressing organization can address both.
- Success Factors addressing new or improved corporate organization elements or conditions as a result of the Strategic Initiative are addressing the Strategic Initiative solution, not the Strategic Initiative process.
- Whenever a Success Factor addresses the quality of corporate processes, corporate solutions and products, or corporate organization attributes to be achieved because of the Strategic Initiative, we talk about a Success Factor that addresses the Strategic Initiative solution.
- A Success Factor can address solution, process, and organization at the same time.

Even if you get no questions about the solution-oriented Success Factors, you can explain that the solution fitting the needs of the Strategic Initiative can be very complex, for example:

- A merged organization
- An acquisition
- A factory
- An information system
- A new subsidiary

The solution can even be a combination of these elements. The participants will decide on what the solution is constrained by the initial scope of the Strategic Initiative.

When we performed PQA in the organization that developed and implemented the DANCOIN cash card, the solution was not restricted before the participants had formulated their visions.

One vision was that the cash card completely replaced the wallet of the owner. You can imagine the cash card with club memberships, credit cards, social security ID, etc.

Fortunately enough, the sponsor was there and could state very clearly that the objective was to produce a cash card with anonymous cash to buy newspapers, clothes washing, soft drinks, bus tickets, telephone calls, and other simple consumption—and nothing else.

Obviously, this restriction could have been stated in the scope in the introduction to PQA, but we had not foreseen such an interpretation of the cash card. It is quite important to state what is not included in the scope if this is not obvious.

The discussion led to other interesting solution observations that in the end allowed the team to define all elements of the solution comprising:

- Communication and networking
- Card distribution
- Card usage
- Card reader
- Central transaction clearing

The "simple" cash card solution was complex enough to challenge all PQA team members.

When you explain the process object, you can use the coffee bean methods to explain the different processes that are required in order to develop, implement, and quality/project manage the Strategic Initiative result delivery by using the graphic representation shown in Figure 3.5.

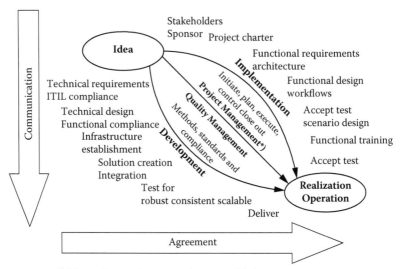

FIGURE 3.5
Process components.

Quality Management and Project Management synchronize Implementation and Development in such a way that the produced solution components, organizational elements, and documentation obtain an agreed quality that complies with the scope that the participants will define and agree to under PQA.

You can show a set of questions that the PQA participants might ask themselves while they define their vision, mission, and suggested Success Factors with respect to the Strategic Initiative (Figure 3.6).

How are reliability, maintainability and availability defined?
What other systems do we integrate with?
Who is the user of the product?
How should the IT infrastructure communicate with the users?
User involvement?
The implementation process?
Use of method, techniques and tools?
What standards do we have to develop and/or implement?
Education, training and coaching requirements?
What are the most common recovery problems encountered
    and what are their prevention requirements?

FIGURE 3.6
PQA questions.

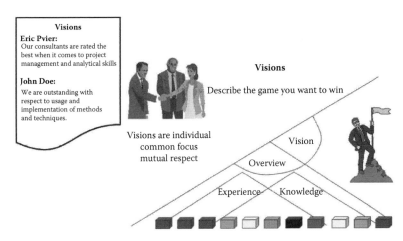

FIGURE 3.7
PQA participant visions.

3.8.3 Definition of Personal Visions and Missions

A mission statement describes what you want to do now in order to contribute to make the vision come true.

A vision statement describes what you want or expect the situation to be in the future when the Strategic Initiative or the task has completed (Figure 3.7).

Major corporations often show their vision and mission statements on their public web sites. I have chosen to show examples from Volvo and Novo Nordisk:

Mission and Vision of the Volvo Car Group (from its public web site)
Vision
To be the world's most progressive and desired luxury car brand.

Mission
Our global success will be driven by making life less complicated for people, while strengthening our commitment to safety and the environment.

The Novo Nordisk Mission and Vision are expressed in The Novo Nordisk Way Essentials:
The Essentials are ten statements describing what the Novo Nordisk Way looks like in practice. They are meant as a help to our managers and employees for evaluating to what extent our organization acts in accordance with the Novo Nordisk Way.

Mission and Vision of the Volvo Car Group (from its public web site)
Vision
To be the world's most progressive and desired luxury car brand.

Mission

Our global success will be driven by making life less complicated for people, while strengthening our commitment to safety and the environment.

The Novo Nordisk Mission and Vision are expressed in The Novo Nordisk Way Essentials:

The Essentials are ten statements describing what the Novo Nordisk Way looks like in practice. They are meant as a help to our managers and employees for evaluating to what extent our organization acts in accordance with the Novo Nordisk Way.

The Essentials are as such an important means for identifying actions, which our organization may take to further align our way of working with the thinking, and values that characterize the Novo Nordisk Way.

- We create value by having a patient centered business approach.
- We set ambitious goals and strive for excellence.
- We are accountable for our financial, environmental and social performance.
- We provide innovation to the benefit of our stakeholders.
- We build and maintain good relations with our key stakeholders.
- We treat everyone with respect.
- We focus on personal performance and development.
- We have a healthy and engaging working environment.
- We optimize the way we work and strive for simplicity.
- We never compromise on quality and business ethics.

Corporate vision and mission statements are developed by corporate leadership to indicate to employees and stakeholders in general where they want the corporation to go and how they intent to arrive at this target—their mission.

Under PQA, we do not develop common vision and mission statements. We leave that role to corporate leadership, who might use some other brainstorming techniques to arrive at corporate vision and mission statements.

Under PQA, the vision and mission statements are personal. Their purpose is to present the participants to each other to obtain mutual understanding and respect.

As can be seen in the previous examples, there is a tendency to mix up vision and mission, and in the PQA case, you are welcome to do the same.

Good vision statements will have some of these features:

- Easy to understand by other participants
- "Paint" a picture of future values

- Express something that you want to happen
- Might be SMART (Specific, Measurable, Achievable, Realistic, Time bound)
- Aligned with corporate strategy and the scope of the Strategic Initiative
- Address the purpose of the task at hand
- Describe the relationship to stakeholders
- Describe what you want to do and achieve

Each PQA Workshop participant defines a personal vision and mission. The participant presents a personal view on the expected result and the expected business benefits.

Each individual vision is written down with the name of the author attached to it. It is very important for the cooperation and mutual respect among the team members that they understand the motivating factors of each other.

It is allowed and even recommended to ask for explanations of the vision and mission statements, but their relevance or correctness can never be challenged because it is a personal choice.

People tackle problems in different ways. We like to know what the attitudes are among the members of the PQA Team. Based on this knowledge, we can play better on each participant's strengths and motivation in order to obtain maximum creativity and synergy.

3.8.4 Definition of Success Factors with Quality Control

The Success Factors express what the PQA team thinks should be the quality of the result of the Strategic Initiative. The result can relate to any solution, process, or organization element that is relevant to the Strategic Initiative or the task (Figure 3.8).

The Success Factors also express what the PQA Team thinks how other people will evaluate the project result.

The Success Factors express opportunity events and conditions or they express that a threat event or condition is avoided, but not how this is done.

To perform an activity cannot be a Success Factor.

You cannot ensure that the result of an activity contributes positively to the value of the Strategic Initiative unless we know what that contribution is, that is, it is the event or situation of this contribution that is the Success Factor, not the activity.

Needed activities are defined after that the Success Factors have been established in response to these, not the other way around.

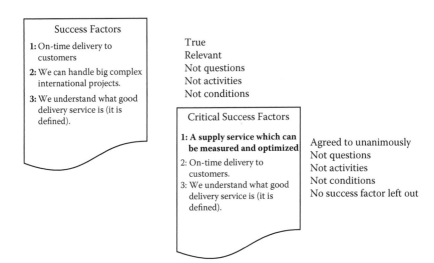

FIGURE 3.8
Success Factors and Critical Success Factors.

Questions can of course not be Success Factors, but good questions might direct the participants to think about other SMART Success Factors.

The Success Factors are identified by letting each participant in turn suggest one Success Factor. All suggestions must be true, relevant, and valid and, at best, SMART. Full unanimous agreement on each suggested success factor is not required, but the truth, relevance, and validity must be ensured by the team—sometimes simply by reformulation of the suggested success factor.

The process of Success Factor suggestion is continued until no team member has any more suggestions.

All the suggested Success Factors are listed on flipcharts, numbered sequentially without any priority, and visible to all team members once there is agreement about the formulation of each one (Figure 3.9).

When a flipchart has been filled in, it is glued to the wall of the meeting room so that all flipchart sheets and success factor suggestions are visible all the time. In this way, you can avoid too much duplication of suggestions, but to avoid duplication as such is not necessary. A Success Factor suggestion can only be doomed to be duplication if the suggesting participants agree to this.

As facilitator, you are allowed to suggest a Success Factor formulation, but not to change the idea of the suggesting team member. Other team members are allowed to express their opinion and to try to change the idea of the suggesting team member.

> **Suggested Success Factors (Cont.)**
>
> Ⓢ12. The system should be easier, faster, and smarter to use than a paper-based system.
> Ⓢ13. All healthcare service data must be contained in an electronic medium.
> Ⓢ14. Well-defined interface to De Mars.
> Ⓢ15. A flexible system that can be quickly adapted to changing requirements.
> Ⓢ16. FOSIS must for each information user be able to present the necessary information in an understandable and clear manner.
> Ⓢ17. FOSIS must be able to communicate with civilian healthcare systems.
> Ⓢ18. The tender must be completed by July 1.
> ⓄⓈ19. The system must be able to communicate to peer users and senior authorities.

FIGURE 3.9
Flipchart sheet with evaluated suggested Success Factors.

A team member can suggest one and only one Success Factor each time it is the turn of this team member to suggest a Success Factor, even if the team member has a whole list of suggestions. This is the reason for allowing team members to steal suggestions from other team members who have not had the patience to wait for their turn.

The bigger the PQA Team the more Success Factor suggestions you will get. A good team size of 7 to 9 participants will yield 40 to 60 Success Factor suggestions.

Once no team member can add an additional Success Factor, you as the facilitator evaluate the results after sending the participants out of the War Room for a lunch or coffee break (Figure 3.10).

For each suggested Success Factor you put one or more circled evaluations on it that declares if the suggested success factor concerns solution©, process℗, organization©, or any combination of these. For example:

1. More effective Annual Work program preparation©℗©
2. Improved work program reporting℗

FIGURE 3.10
Success Factor evaluation criteria.

3. Knowledge that can be used⊚
4. Application Owners can verify their budget and budget performance⊚
5. Improved agreed business processes supported by the tool⊚⊚

You then count the Os, the Ps, and the Ss:

⊚: 2

⊚: 3

⊚: 3

If any O, P, or S comes out with zero or comparatively few suggested Success Factors, you demand the PQA team to suggest Success Factors concerned with these once they get back from their break.

During early phases of a Strategic Initiative, the PQA teams are normally more solution focused than organization and process focused, and the contrary is the case in later phases.

As PQA facilitator, you are patient when the participants discuss the quality of suggested Success Factors. You are not in a hurry and the discussions are important for an open and positive cooperation between participants, so you only coach with suggestions when you have a good idea for a formulation. You should provoke the PQA Team to suggest a number of Success Factors in all categories if necessary after your quality control.

Concerning the PQA Workshop schedule, you are more careful with time control if you do the whole workshop in one day than if you have two days. Normally, the participants will finish the definition of Success Factor suggestions in 3 to 4 hours and the decisions about Critical Success Factors in 2 to 3 hours. However, if you have more than eight participants, it can take longer. A two-day workshop allows for fun and the second day is normally dedicated to critical success factor definition, activity definition, and quality control of the workshop result.

3.8.5 Definition of Critical Success Factors

The definition of Critical Success Factors is a decision-making exercise. Until this session, the team has only had relatively innocent discussions about visions, missions, and Success Factors. Now the team must reach unanimous decisions about how to classify the Success Factors into Critical Success Factors and about how to formulate the Critical Success Factors.

In this process, the team needs your experience and coaching most. Use your quality control classification into solution, process, and organization-oriented Success Factors to find some obvious candidates for the first class to be defined, for example, two solution-based Success Factor suggestions.

Try to make the first class easy to establish and do not hesitate to suggest a formulation that the team members can improve among themselves, that is:

"We get an optimal production environment."

Such a Critical Success Factor is usable for classification, but it is not SMART because it is not specific.

Ask the participants to suggest a formulation that is SMART.

It might be that the suggested Success Factors that belong to a less SMART Critical Success Factor together make it SMART, but that also indicates that a better formulation of the Critical Success Factors might be possible.

You might have tried out classification based on yellow notes, where each person has the opportunity in a round robin process to group and regroup the notes in silence until no more regrouping is wanted or needed. After this process, you baptize each group to become your Critical Success Factors.

I do not like this method because it does not give good discussions until it is too late after the Success Factor groups have been established in silence. You do get a discussion about formulation, but not about why the Success Factor groups are established the way they are. In addition, you normally do not use a suggested Success Factor more than once, which is not realistic.

The preferred classification method allows the PQA team to discuss each group of suggested Success Factors group by group and formulate the Critical Success Factors during this discussion.

Once a Critical Success Factor has been formulated this way, you easily find other suggested Success Factors to put into the group. Once a suggested Success Factor belongs to a group, you write the group number besides the suggested success factor (Figure 3.11)

Any suggested Success Factor could belong to more than one Critical Success Factor.

The definition of Critical Success Factors close out when all suggested Success Factors belong to at least one Critical Success Factor class.

List the Critical Success Factors on one or two flipchart sheets. Give a number to each Critical Success Factor, but the numbers do not in any way prioritize the Critical Success Factors (Figure 3.12).

Give a number to each Critical Success Factor from 1 to N, where N normally is between 3 and 9.

Suggested Success Factors (Cont.)

②12. The system should be easier, faster, and smarter to use than a paper-based system.
⑨13. All healthcare service data must be contained in an electronic medium.
⑤14. Well-defined interface to De Mars.
⑥15. A flexible system that can be quickly adapted to changing requirements.
②16. FOSIS must for each information user be able to present the necessary information in an understandable and clear manner.
⑤17. FOSIS must be able to communicate with civilian healthcare systems.
①18. The tender must be completed by July 1.
⑤19. The system must be able to communicate to peer users and senior authorities.

FIGURE 3.11
Flipchart sheet with Critical Success Factor reference.

The PQA Teams always wonder why there are only from three to nine Critical Success Factors, but this is actually quite logical.

You have three core quality objects that you define for your task at hand or Strategic Initiative, and for each core object you have three attributes that you also want to cover with your Critical Success Factors.

Some Critical Success Factors address more than one attribute, which gives you between 3 and 3 × 3 Critical Success Factors.

If a Critical Success Factor addresses more than one core quality object, I would normally look for an opportunity to split it into better formulated Critical Success Factors so that three Critical Success Factors really are the minimum number that you will see.

Critical Success Factors

(1) Essential FOSIS functionality implemented simultaneously on time
(2) Intuitive Danish language user interface
(3) FOSIS supports all healthcare services throughout
(4) FOSIS provides access to necessary and complete healthcare information
(5) FOSIS communicates with relevant systems
(6) FOSIS is aligned with the defense IT strategy and is based on relevant standards
(7) FOSIS meets all requirements for safety and traceability
(8) FOSIS enables a flexible, user-specific data handling
(9) FOSIS increases quality and efficiency in healthcare service

---

FIGURE 3.12
Flipchart sheet with Critical Success Factors.

The PQA Team agrees 100% to each critical success factor. If someone in the PQA Team finds that an additional suggested Success Factor could improve the specificity of a Critical Success Factor, such a suggested Success Factor can and should be added if the team does not object to it.

We cannot lose one single suggested Success Factor in the classification process, that is, all suggested Success Factors must belong to at least one Critical Success Factor.

3.8.6 The PQA Matrix with Activity Definition

The PQA matrix is used for definition of required and adequate activities to ensure the fulfillment of all suggested Success Factors and Critical Success Factors.

The Critical Success Factors are used for control of completeness and outline scope of the activities that the PQA team defines.

Just before the session of Activity Definition on the PQA Workshop, you as facilitator draw up the PQA Matrix on one or two flipchart sheets—the number of flipchart sheets is not important, you can add more if you define many activities (Figure 3.13).

The PQA Matrix is drawn up with a number of columns corresponding to the number of Critical Success Factors plus three columns:

- The first column for the Activity Name
- Each of the following for a Critical Success Factor referenced by its ID number
- The next two for value and responsibility

PQA Matrix											
Activity Name	1	2	3	4	5	6	7	8	9	Value	Who

FIGURE 3.13
PQA Matrix flipchart layout.

The PQA Matrix is written on one or more flipcharts. The rows represent suggested activities described with a simple headline, such as the Activity Name.

For each activity, you indicate with a star or some other symbol that this activity will contribute to the fulfillment of the referenced Critical Success Factor.

If the outline scope is not defined precisely enough by the set of Critical Success Factors referenced by the activity and the underlying suggested Success Factors, then you can add a supplementary comment to the Activity Name.

The activity leader is appointed among the PQA Workshop participants and the initials of the activity leader are added in the Who column.

Each participant can suggest as many activities as the participant finds relevant. As a facilitator, you coach the team to get activities defined on an equal level (e.g., new projects, activities to be further broken down into tasks, or tasks on their own).

As facilitator, you are also allowed to suggest activities.

The PQA Matrix document can look like this:

Critical Success Factors						Value in execution	Who is Responsible
5. Quality management							
4. The system supports a dynamic business environment							
3. Accessible information							
2. A competent organization							
1. A supply service which can be measured and optimized							
Activities	1	2	3	4	5		
Define all aspects of a good supply service	*	*			*	4	LH
Build a development support organization		*			*	2	JD
Build the user competence necessary to utilize the new system		*	*	*	*	3	LH
Define the user competence necessary to define the new system	*	*	*	*	*	2	JD
Distribute the IRS report		*		*	*	1	PP
Inform involved sales companies about the process and the progress ongoing		*			*	1	PP
Define and build the complete system for communication, HW, SW and applications	*	*		*	*	0	CV
Do PQA with the users involved in the design	*				*	0	CV

The participants control the scope of each activity by entering a * under each Critical Success Factor that will be addressed by the activity.

Do not enter * where only indirect relations between Critical Success Factor and activity exist because this might lead to * in all cross-reference points.

If you can enter * under all Critical Success Factors, it is a sign that the activities have been defined on too high a level.

Once all the * connections have been defined, you check that the activities that address a Critical Success Factor all together can fulfill the Critical Success Factor and the underlying suggested Success Factors.

If some activity is missing, it is added and new * made for cross-reference between this new activity and eventual other Critical Success Factors.

You continue until all success factors are fulfilled by the outlined activities.

You then evaluate if an activity is already executing. If that is the case, you give it a value between 1 and 5 that indicates the following:

0   Activity not done
1–2   Activity executing but unstructured performance
3–4   Activity executing with a known manager, but to be improved after PQA
5   Activity executing well with a known manager

The PQA Matrix activity definition session is closed out for each defined activity with the assignment of one of the participants in the PQA Workshop as activity leader.

The activity leader plans the activity in detail after the workshop and suggests a manager of the activity if such a manager does not already execute the activity.

3.8.7 PQA Activity Description and Assignment of an Activity Manager

On the PQA workshop, you present and explain again how to use the Standard Activity Description before you close the workshop out with review planning.

The PQA team members that have been appointed as leaders of one or more activities define the activities in depth after the workshop using the Standard Activity Description form.

Activity Description, selection and suggestion of resources and appointment of an activity manager are classic project management activities. It is not the purpose of this book or this chapter to explain these in general.

Nonetheless, the content and the quality expectations concerned with the Standard Activity Description are covered next because it is part of PQA.

3.8.8 PQA Workshop Review Planning in the PQA Workshop

The further planning of the Strategic Initiative is done in PQA Workshop review meetings. The PQA Workshop review meetings are new workshops that build on the result from the initial PQA Workshop and previous review meetings.

All participants in the initial PQA Workshop participate in the first PQA review in order to ensure that agreed activities are complete, valid, and consistent to succeed with the Strategic Initiative. It is also important to ensure that suggested deadlines, resource usage, and budgets are realistic and that the suggested results and deliverables are achievable in light of resource availability and other concurrent activities.

It might be difficult to keep a PQA Team on very high strategic level together for PQA reviews, but if you succeed in doing it, you can ensure the highest possible probability to succeed with the Strategic Initiative.

When completeness of the PQA Team cannot be achieved for a review, cancel or postpone the review if one of the unavailable resources is necessary for decision making. The other resources will simply waste their time that could have been better used on other activity.

You can compare the situation of an incomplete PQA Team with the situation of key resources missing for producing an important project deliverable on a project task. If you pursue the task, the risk is high that the work is interrupted or that the result is unacceptable, which inevitably leads to delays and increased cost. In most cases, you do not want to take that chance. It is highly recommended to perform tasks only when required resources are available. This is true for material, infrastructure elements, and human resource resources.

Imagine what happens if you start programming a web-based solution and you have no web access—this would not bring you very far unless you have good test and simulation facilities available. Even with such development facilities, you might get surprises when you execute the program on the real web. Furthermore, if your preferred web programmer who is used to agile development is not available, you will waste a lot of time if you bring in somebody who might be competent but who does not know you or your requirements.

Think twice before you bring in new resources in order to speed up things—most often new resources bring more chaos than progress. If you need additional resources, make sure to plan for it and to allow for them to be competent and to be adapted to your activity. This applies to people and technology.

PQA Workshop review workshops comprise the following types:

- The simplest review workshop type is always done. It verifies that the Activity Descriptions that have been agreed to be developed are satisfactory as a requirements specification for the work to be performed in order to deliver the needed result of the Strategic Initiative. It is verified that each activity does what is required from that activity and that all the activities together deliver the expected result of the Strategic Initiative without doing work twice, that is, that the activity integrity is ensured.
- The more complex form of PQA Workshop review is a risk management based review. Once the task of the simplest form of review has been performed, the Strategic Initiative risk situation is evaluated based on the detailed risk events and conditions such as those that have been described in each Activity Description.
- The most complex form of PQA Workshop review is a planning workshop where the complete project plan for the Strategic Initiative is assembled, evaluated for resource availability, scheduled, approved, and signed off by the PQA Team for further sign off by the sponsor and sometimes by the original sponsor.

In the initial PQA Workshop, you and your sponsor explain these review types and the importance of keeping the PQA Team together until the complete project plan has been signed off.

It is important that the development of Activity Descriptions be performed as fast as possible after the initial PQA Workshop.

It is important that all agreed Activity Descriptions have been completed and distributed to all PQA Workshop participants at least one week before the PQA Workshop review.

The participants who are responsible for the development of the Activity Descriptions get the time they ask for to verify directly with the required and competent resources to be involved with the delivery of the task result:

- That the resources agree with the scope and the solution quality to be delivered
- That the resources can perform the work
- When the resources are available to do the work
- The working conditions the resources demand to be able to do the work

It is important for the scheduling of PQA review workshops that the preparatory work of Activity Description development can be done with good quality.

With this constraint in mind, the PQA review workshops should be done as fast as possible.

The PQA facilitator works with the Project Office or a similar secretarial function to support the Activity Description work and ensure that Activity Descriptions comply with the standard (a standard is described next).

The PQA participants will continue to be a team and they will normally help each other to get the Activity Descriptions done in time and with good quality.

The direct PQA Workshop result is printed and delivered to the participants within two working days after the workshop.

3.9 THE ACTIVITY DESCRIPTION PRODUCTION

The PQA participants who are responsible for the development of the Activity Descriptions have been chosen because they understand the scope of the activity and because they are motivated to coach the planning and execution of the activity, not because they know in detail how the activity can be performed or what the quality of the result can be.

These participants can be the project managers of the activity or they can suggest someone outside the PQA Team to be the project manager of the activity.

Whatever the case, the responsible PQA Team member and the future activity project manager get together with the future Workgroup Team that is responsible for performing the activity and delivering the result of this activity.

The Workgroup Team assists in the production of the Activity Description to be reviewed by the PQA Team in order to ensure valid and reliable estimates and quality expectation.

In some cases, an activity is a big project with many stakeholders and many resources to produce a complex result. In this case, a PQA Workshop introduction to PQA with the Workgroup Team can replace the Activity Description. The activity becomes a sub-project to the Strategic Initiative. In major Strategic Initiatives with several sub-projects, the planning process will take longer.

If the Strategic Initiative involves many external organizations and internal departments, it is investigated if it can be executed as a program with a Program Office.

The Program Office can ensure legal and contractual compliance, tracking, and efficient communication in the governance of Strategic Initiatives facing high complexity of organization and solution.

In all other cases, the Workgroup Team will produce the Activity Description and be committed to the estimates and the scope of the activity such as they have expressed it themselves.

3.9.1 The Activity Description Preparation

The Activity Description is done for each activity from the PQA matrix as well as the ones that will become a new project that requires its own PQA.

For the preparation of the Activity Description, use the PQA result and the PQA Matrix:

Critical Success Factors:							EXECUTION VALUE	RESPONSIBLE
6) Conformity between LI services and WCAT expectations								
5) The solution contributes measurably to increased profitability								
4) Applied and accepted management tool								
3) Conformity between our delivered services and the customer expectations								
2) The system supports optimal project implementation								
1) Effective project culture								
Activities:	**1**	**2**	**3**	**4**	**5**	**6**		
1) Establish project team	*	*		*		*	1	MI
2) Establish project office	*	*	*	*			0	HK
3) Define and document the essential project types		*	*		*		2	ML

In the following example, ML writes the Activity Description for:

3) Define and document the essential project types.

To this end, ML has called the future Workgroup Team together to help with the scope setting, the estimation, and the risk identification.

ML and the Workgroup Team will use the PQA Workshop result that is relevant for this activity to ensure compliance of the Activity Description with all referenced success factors.

2. The system supports optimal project implementation

2	Reliable data that are validated before they come into the system
7	The system is integrated with other information systems, so that the information is coordinated across these systems
8	Visible utilization of resources (billable time versus time spent)
9	The system must make it easier for the users to do planning
17	Utilization of the system gives a measurable economic benefit
19	We get an overview over where we lose and make money, so that we get an improved business focus
22	The invoice foundation appears significantly faster
23	Increased maneuverability
24	Flexible reporting capabilities that show relevant information
25	Each employee has an overview of the tasks that the employee is allocated to and used on
26	Be able to identify potential conflicts and deviations early in the project
33	Bottlenecks are visible
34	Internal and external plans can be maintained in the same place
37	The system is easily adaptable to new methods at WCAT
39	It will be visible if a task is behind or ahead of schedule
45	The system is proactive—it reminds the user about activity that must be initiated
47	No excuse for not being proactive and for not taking initiative
50	We can assess the impact of different projects and project portfolio scenarios
56	The system contains only one truth
57	We get fast and useful final costing of projects
59	The system can highlight the vulnerability relative to essential staffing
65	You can register all kinds of time spent in the system

3. Conformity between our delivered services and the customer expectations

10	Controlled project process with fewer surprises—higher predictability
16	The planned project times are respected
20	At any time we can inform the customer about the status of the costs incurred in the customer's projects
23	Increased maneuverability
28	Higher customer satisfaction because we deliver what is agreed on time

29   Visible customer deliveries and consequences of the customer's failure to comply with agreed conditions

31   The system provides a better basis for guiding the customer to an optimal process

36   Better time estimation for proposals (standard time)

40   It becomes apparent who must be notified when there are deviations from a plan

42   All involved stakeholders report problem situations in the system with the assurance that the situations are treated in time

48   We must be able to detect and measure the quality of each process (e.g., agreed with the customer)

5   The solution contributes measurably to increased profitability

8   Visible utilization of resources (billable time versus time spent)

11   Increased reusability of collected data and experience

17   Utilization of the system gives a measurable economic benefit

19   We get an overview over where we lose and make money, so that we get an improved business focus

22   The invoice foundation appears significantly faster

36   Better time estimation for proposals (standard time)

55   We start up with a solution that quickly provides visible benefits and is widely used

57   We get fast and useful final costing of projects

3.9.2 The Activity Description Guideline

Activity Description	99) Activity name as this is printed in the PQA Matrix. The number is the activity number from the PQA Matrix. When new activities are established during the PQA Review process in response to risk or for other reasons, such activities get a responsible PQA Team member assigned and an Activity Description is produced by the Workgroup Team to perform the activity and deliver the result.
By: The PQA Team member who is responsible for the Activity Description	Delivery date: The date the Activity Description was delivered for review. Approval date: The date the Activity Description was signed off.
Scope	Describe WHY this activity is required and what its areas of concern and responsibilities are in the context of the Strategic Initiative. Also, describe what the activity does not concern if this can be discussed. If possible, make direct reference to success factor expressions.

Deliverables	A description of the expected outcome, for example, a tender material, a report, or an accepted system. Supportive documentation is referenced here.	
Purpose	A description of the deliverable quality expectations or of the expected benefits from the delivery of the deliverable. If compliance is required for legal, contractual, quality, or other types of constraints, it is mentioned here why this is the case. Here you can also say that the activity ensures fulfillment of "success factor expression." Of course, it is implicitly ensured that the activity contributes to fulfillment of related success factors.	
Responsible	The person responsible for getting the activity done (sometimes the person writing this description). This is the activity Project Manager. In some cases, the Project Manager can be a vacant position. This is not an optimal situation because it means that the Activity Description does not have commitment from a Project Manager before the PQA Team approves it. Try to avoid this situation.	
Resources/Roles	A description of needed roles and their responsibilities and skill and competence requirements. Specific named resources can be applied to the roles, but you should not put a person in here without mentioning the roles that the person will take on during the performance of the activity.	

Task List	**Task Description**	**Estimated Resource Usage**
	Scope, purpose, and deliverable for each task executed in this activity. Describe why you need the roles that have their usage estimated concerned with this task. ID) Task name as shown in project plan Scope, purpose, and deliverable (short, because this will be defined by the Project Manager or the performing resources during result production).	Roles/names of resources to perform task with person-hour estimate for each resource; also the non-human ones if relevant.
Timeframe	Duration in days, weeks, or fixed period (between this date and that date).	

Risk Assessment	Describe potential events or task conditions that could influence the deliverable quality, duration, or estimated resource usage.
	Think about external factors, which cannot be controlled, and internal factors, which can be controlled by project management.
	Describe the probability of the event or condition (if probability is 100%, then it is a problem and not a risk). Describe the impact of the event.
	Formulation: "This xx event might happen with high probability which will increase yy resource usage with a factor of 1.5."
	The complete risk situation of the Strategic Initiative will be listed in the risk response matrix where the risk response strategy is also shown.
	Leave the task specific risks here even if they are also shown in the risk response matrix.
Dependencies	Reference to activities that are performed before this activity is performed, concurrent with this activity, and to activities that will use results from this activity.
	Predecessor activities:
	ID)aaaa
	Concurrent activities:
	ID)bbbb
	Successor activities:
	ID)cccc
	Explain why the relationship is relevant if not obvious.

While preparing the Activity Description, you ensure that all Critical Success Factors and their related Success Factors from the PQA Matrix implicitly or at best explicitly are addressed by the Activity Description.

You can use the scope description and the purpose description of the Activity Description to make sure that the activity deliverables explicitly contribute to the related Success Factors from the PQA Matrix.

3.9.3 Activity Description Example

ML and his Workgroup Team reached the following result, which was presented for review in the PQA Team review meeting after approval by the Facilitator and the Project Office for compliance with the Activity Description guideline.

Activity Description	3) Define and Document Essential Project Types
Prepared by ML	Preparation date: 04.06 PQA Team approval date:
Delimitation	This activity does not include the following: The specific project model and the contents of the activities in the individual departments. The organizing of the projects and the roles in the WCAT—project organization).
Products	A document that describes the essential project types with clearly defined phases, when to change phase, and the distribution of responsibilities. During this, we should determine the interfaces and dependencies between the various functional areas.
Purpose	The purpose of describing the essential project types is to define and document these in a structured way, and from this to: • Be able to prioritize based on WCAT business objectives. • At any time, be able to see in which phase a project is and who is responsible. • Ensure that the individual employee has an overall view of the phases in essential project types. • Ensure a structured collection of experience-figures for each project type. • Ensure that the descriptions form the basis of the implementation in the COTS.
Responsible	JO
Other resources	CT; ML Qualifications: Thorough understanding of WCAT business processes (service, workflow, and products)

Sub-activities	Description of Tasks	Resource Requirement
	1) Identification of the essential project types.	JO, ML (4 hours)
	2) To carry out the analyses and describe the individual project types a group (person) per project type is appointed. It is the responsibility of the individual group to: • Describe the existing project workflow • Identify the action areas/problems • Determine and describe the ideal project workflow (phases, when to change phase, etc.) • Describe the distribution of responsibilities for the project workflow	Expected time consumption per group: 2 weeks
	3) Configure the COTS system for the essential project types.	CT, LI, ML, JO Expected time consumption: 4 weeks
Time frame	August – October	
Risks	If this project is *not* given top priority by the management, the resources will disappear from the project. A strong project team must be set up to ensure its visibility in the organization. Efficient project culture does *not* arise by itself. The COTS facilities are *not* able to support our description of the project types. The system is *not* easily adapted to new project types.	
Dependency on other activities	This activity can be started independently of other activities. Close coordination (concurrent) with: 3) Define and document effective project management	

This Activity Description is not perfect. The estimates could be more precise and the risks could be better formulated, but as Coach and Facilitator, you should not be too critical with the PQA Team performance. You should expect to accept even less perfect Activity Descriptions for review.

Missing elements, consistency, more precise estimates, and risk formulation can and should be handled in the PQA Review Workshop, where you can share your knowledge and experience with the PQA Team.

3.10 THE PQA REVIEWS

The PQA Workshop reviews are carefully prepared by the Facilitator and supported by the Project Office before the review meetings. Activity Descriptions have been approved to comply with standards and they have been distributed with reasonably good quality to all participants at least a week before the review workshop.

The Project Office ensures availability of needed facilities, the required PQA Team members, and the timely distribution of needed documentation to all PQA Team members before review meetings.

You need the Project Office to ensure that the PQA review participants have received the Activity Descriptions in time.

Although review dates and participation have been agreed on in the PQA Workshop, you will meet events and conditions that require changes to the review dates:

- Participants who are committed to perform other activities for valid private or professional reasons
- Activity Descriptions that are more complicated to develop than expected and will require more time to reach an acceptable quality
- Future required Workgroup Team members are not ready to help the PQA Team member to develop the Activity Description, which will be delayed for that reason

Changing review date is hard work that requires a good secretarial Project Office function.

The preparatory work is best performed with access to efficient planning and tracking tools managed and used under the support of the Project Office. With such tools, the estimates become more realistic and the Project Office can prepare for PQA Team planning by:

- Creating the Strategic Initiative as a project in the corporate project database

- Making sure that suggested resources exist in the corporate project database
- Making sure that suggested resources can be used for planning, that is, they are allocated to the Strategic Initiative
- Creating the project plan activities as they are defined in the Activity Descriptions leaving the definition of milestones and phases to the PQA Team decision making
- Creating dependencies between activities as these are defined in the Activity Descriptions
- Suggesting a schedule based on the estimates in the Activity Descriptions and already produce the first warnings of non-availability of resources to perform the activities as wished
- If the corporate project management system comprises risk management, the Project Office can already add the suggested risks to the risk list of the Strategic Initiative project plan.

The Project Office tools and preparation do not replace the PQA Team brainwork, but the Project Office preparatory work saves a lot of time for the Facilitator and the PQA Team during the review, especially if the review is performed in a War Room with a big white board screen connected to the corporate project management system.

- Resource usage conflicts are immediately visible.
- Resource availability to activities is immediately visible.
- Complete activity overview and total schedule is visible and can be played around with.
- Alternative resources can be discussed for Project Manager roles and Workgroup participation.
- Initial risk overview is established and duplicates can be identified.
- Risk formulations can be corrected directly for agreement across the PQA team.
- Risk responses can be related directly to activities.

The Project Office is an advanced secretarial function that looks after how established standards for project management are used in and across projects in the organization.

This means that the Project Office can perform the first Activity Description review simply to control that it complies with the standard form and can be used for creation of the project plan and schedule as a foundation for the first Strategic Initiative baseline.

If you do not have access to Project Office support, you are obliged to establish this support, even if you have to create your own Project Office.

The PQA Team members are able to make high-level pertinent planning decisions based on the delivered PQA documentation, status reporting, Activity Descriptions, and already produced plans and schedules. However, if the PQA Team members cannot see the full consequences of their decisions by using an appropriate Project Office supported planning system, then they are working in the dark and the resulting plan is a waste of time.

3.10.1 The Basic Activity Description Review Workshop

The general purpose of the basic PQA Workshop reviews is to ensure that:

- Suggested activities and their deliverables are within the scope of the Strategic Initiative
- The set of activities is complete with respect to delivery of the expected result of the Strategic Initiative
- Activities do not make the same things more than once (integrity check)
- Activities are executed in the right sequence
- Resources are available for fast and efficient activity execution as estimated
- Effective communication is established in activities, between activities, and between corporate management and activity management

The Activity Descriptions and the PQA Team brainwork will not be enough to reach a consistent plan and schedule that can fulfill all the success factors. You need access to a well-prepared project setup in a project management support system that allows the PQA Team to simulate the consequences of its decisions.

If the PQA Team or the facilitator does not have experience with using the corporate project management support system, they need support from the Project Office during the PQA Workshop review.

3.10.2 The Risk Management-Based Review Workshop and the Risk List

Where the initial PQA Workshop focused on opportunities (Success Factors), the risk management process during a PQA Workshop review focuses on threats and threat response with the objective to establish the most optimal project plan for the Strategic Initiative.

Risk Management with identification, evaluation, and documentation of threats and opportunities and definition and documentation of the response strategy can be performed in a group dynamic process during a PQA Workshop review.

The threats might be classified analogous to the CSF (e.g., Critical Threats), which will make it easier to control the completeness of the responses to the risks.

The risk management-based PQA Workshop review can be done at the same time as basic PQA Workshop reviews if the risk situation is not too complex.

The purpose of the risk management-based PQA Workshop review is to identify the risks, to evaluate the risks, and to respond to the risks:

- Risks are identified with correct formulation (e.g., "If aaa event or bbb condition happens, then consequence xxx will be the result")
- Risks are evaluated and quantified (exposure = impact value × probability of event or condition)
- A risk response strategy is formulated that can comprise:
  - New activities
  - Activity integration
  - Improved activity performance
- Improved Activity Descriptions explicitly respond to identified risk events and conditions

The identified risks with their responses are documented in the Risk Response Matrix for follow up with the PQA Team and for communication to other stakeholders.

The Risk Response Matrix is used for control of the completeness of the risk response strategy established while the PQA participants are reviewing the Activity Descriptions on the risk-oriented PQA Workshop review.

The Project Manager or the PQA Facilitator produces the risk list (Figure 3.14).

The PQA participants ensure the best possible response strategy by controlling that the set of responses to a risk is the best possible way to avoid, mitigate, transfer, or share that risk, and by controlling that the full set of impact on all risks by a given response is fully understood and documented, for example, in the concerned Activity Descriptions.

Project ID	Project Team Aaaa Bbbbb Ccccc	Date Approved	Version ID	Approved by: Aaaa Bbbbb Ccccc	Distribution list Bbbb Eeee Feee	
Response			RISK Exposure	ID) Resources not available in time	ID) If aaa occurs, then bbb fails	ID) ...
				Event or condition probability Impact value	50% $200K	30% $1,000K
Response Activity	Responsible	Deadline	WBS ID	m.m.m of activity concerned	2.1.2; 3.1.4	m.m.m
Procure sub-contractors	LD	22/22/13	n.n.n of activities responding	*		
Hire Java Ace			1.3.2; 2.2.1	*	*	
...			n.n.n			This is a comment

- The first row shows the name of the project (the Strategic Initiative), the version of the list, who has approved the list, the PQA (project) Team, and the distribution list.
- The second row shows the risk names—one risk per column with ID.
- The third row shows the risk exposure.
- The fourth row shows the headers of the first four columns; in the following columns are shown the Work Breakdown Structure (WBS) ID of the activities that are concerned with the effects of the risk in that column.
- The following rows contain in the first four columns:
  1) The description of the risk response.
  2) The person who is responsible for the risk response being executed as decided.
  3) The deadline for the risk response action.
  4) The WBS ID of the activities that execute or are concerned with the risk response.
- In the next columns, you cross-reference risks and risk responses. You can simply indicate a * of reference if the relationship between the risk and the response is self-explanatory or you can explain the relationship with a comment about how the exposure is changed, for example. The exposure change direction can also be indicated by a + or – sign replacing the *.

FIGURE 3.14
Risk Response Matrix.

If you have used a method other than exposure calculation to prioritize the risks, you indicate it in a comment and explain how and by whom it was done. This makes it possible for others to evaluate the result that you have obtained.

To assist you in finding pertinent risks for your Strategic Initiative, you can find a number of standard risk checklists with core risk objects on the web. The risk objects that you would look for depend on the scope of your Strategic Initiative. One such list can be found on http://www.misronet.com/risks.htm:

Technical Risks

- Incomplete design
- Inadequate site investigation
- Uncertainty over the source and availability of materials
- Appropriateness of specifications

Logistical Risks

- Availability of resources—particularly construction equipment, spare parts, fuel, and labor
- Availability of sufficient transportation facilities

Construction Risks

- Uncertain productivity of resources
- Weather and seasonal implications
- Industrial relations problems

Financial Risks

- Inflation
- Availability and fluctuation in foreign exchange
- Delay in payment
- Repatriation of funds
- Local taxes

Political Risks

- Constraints on the availability and employment of expatriate staff
- Customs and import restrictions and procedures
- Difficulties in disposing of plant and equipment
- Insistence on use of local firms and agents

When the scope is implementation of information systems, your checklist could look like this:

Internal risk that can be prevented or mitigated by project management during the conduct of the project can be related to:

- Resource availability
- Stakeholder accept
- Resource quality
- Internal organization
- Safety
- Security
- Confidentiality

External risk that cannot be prevented by project management, but that might be mitigated by project management before the project is initiated can be related to:

- Technology efficiency
- Sub-contractor reliability
- Management attitude
- Funding
- Market
- Legal matters

It is recommended that you establish your own risk checklist. Once established it will be an important source of lessons learned to be maintained by all qualified stakeholders.

A risk checklist saves you a lot of time with Risk Management and it ensures better results from your risk-based PQA Workshop reviews.

3.10.3 The Strategic Initiative Planning Workshop and the Project Schedule

The Strategic Initiative planning workshops are PQA review workshops. The PQA result underlying a Strategic Initiative is together with any documented and approved changes to this result the foundation for the planning workshops.

The Strategic Initiative planning workshops will take place regularly, normally once a month, in order to ensure:

- Communication of approved changes and expected changes
- Approval of suggested changes if the PQA Team is on the Strategy Governance Team level
- Approval of change suggestions if the PQA Team is on the Process Governance Team level
- Plan deviations are discussed and corrective actions are agreed if needed
- Risk events that have occurred are discussed and corrective actions are agreed if needed
- New risk or new risk probabilities are discussed and responses are agreed and documented in improved Activity Descriptions and in the Strategic Initiative Risk Matrix
- Resource availability is ensured for the next period by negotiation between project managers and resource managers
- Rescheduling of activities is agreed in case of non-availability of needed resources or in case of delays of required deliveries from related activities
- Agreement is established between project managers of activities and resource providers

Participants in Strategic Initiative planning workshops comprise all involved management of involved resources. This management also comprises the managers of concurrent Strategic Initiatives or other concurrent projects that demand the resources concerned.

You ensure that sub-contractors are available for planning and scheduling your Strategic Initiative.

The Strategic Initiative planning workshops cannot yield a usable result without a fully up to date project management system database supported by the Project Office in support of the meeting discussions and decision making.

Brainstorming is not a method for Strategic Initiative planning and scheduling. Strategic Initiative planning and scheduling is based on facts and management decisions based on full visibility of these facts.

The facts for Strategic Initiative planning and scheduling decision making are the project tracking statistics produced weekly by the Project Office based on progress reporting from Workgroup Team members and project managers. Project Management and Process Governance Teams comment on the statistics when major deviations from the baseline occur. Deviations from baseline also comprise Strategic Initiative risk situation changes.

If correctly set up by the Project Office and correctly used by Project Managers and project resources, the corporate project management system allows simulation of consequences of decisions made on the planning workshop as a safe foundation for agreement and further progress.

3.11 THE WCAT CASE STUDY

The establishment of improved production methods case in WCAT has already been presented in Chapter 2. Only the result is presented here.

The WCAT case is interesting because the established Workgroups and their projects looked fine originally, but they were not sufficiently strong to survive a sudden change of sponsor.

The new sponsor had ideas that deviated from the PQA result. The new sponsor already knew where to take the organization and the former sponsor's PQA-based strategy did not fit into this picture.

Irrespective of this situation, my organization still delivered its new COTS system and the order project management information system based on this, but our role as Facilitator was finished.

The introduction to the PQA Workshop had the following initial questions:

- What are the 3 to 5 major problems with the current project management strategy?
- Where can your department be more efficient?
- What do you expect from the new information system?
- What do you expect from other departments in WCAT?
- Who must be involved with the specification of the future WCAT project management environment?
- How is the project management environment serviced in operation?
- How is the new WCAT project environment quality managed?

The PQA result is shown in Appendix B.

The thorough selection of participants ensured a high quality result, where improvements to the resulting business processes and the training of the users in the new methodology were in focus to a much higher degree than the installation and operation of the COTS system. Solution, organization, and process are in good balance.

On this basis, we actually achieved an acceptable solution implementation and we know that the enterprise is still very successful.

3.11.1 Activity Description Examples

A selection of the Activity Descriptions produced after the PQA Workshop for the first review meeting is shown in Appendix B.

The formulation of activities to be performed shows here key-stakeholders of high maturity with a lot of experience. This experience allows the stakeholders to see the risks in a broad perspective.

You will rarely find such a level of quality of pertinent Activity Descriptions from a less mature group, and the number of initially identified pertinent risks is impressive.

3.12 THE BANKING INFORMATION SYSTEM SWAP CASE

The program of swapping all IT and Information Systems in a private bank has been presented in previous chapters.

The PQA process used in this case differs from the standard one and allows you to see an example of a deviation from the standard that accelerated progress early, but created quite a few problems later on.

Once the IT Director had established the Strategy Governance Team and the project had been redefined into a program, the Program Manager and the Deputy Program Manager prepared the initiation of the PQA for this program:

- A non-standard PQA Workshop for the Strategy Governance Team to select participants from departmental management level to define Workgroups and Scope of Work for information system requirements specification.
- Preparation of standard forms to be used for Statement of Work documentation and requirements specifications.
- Introduction (PowerPoint) to be used for opening the meeting where the selected departmental managers were demanded to prepare the PQA Activity Descriptions. The invitation to this meeting was a simple e-mail with the PowerPoint presentation attached.

- Preparation of the initial War Room facility that was just the biggest meeting room available in the building with one big oval meeting table ensuring that all participants could see and hear each other.
- Conduction of the special PQA Activity Description Workshop with duration of only one day.
- Documentation of the PQA Workshop result (PowerPoint) with activities to be performed, Workgroups to perform the activities, and required quality of work results.
- Establishment of coaching and facilitation for the Workgroups.

For the following chapters, this case study contains a thorough business analysis with a quite efficient document standard. The resulting requirements specification allowed fast contracting of sub-contractors.

The case study is furthermore interesting in that it demanded a completely new sub-contractor contract to be developed that after often-heavy discussions was signed by all sub-contractors.

The Program Manager supported by the Corporate Legal department and the IT Director developed the new contract.

The sub-contractor contract allowed the program to progress faster with exceptional results from the sub-contractors and the internal bank and IT organization.

The sub-contractor contract is explained in Chapter 4.

3.12.1 The PQA Non-Standard PQA Process

The initial non-standard PQA workshop was conducted as a meeting of the small and dedicated Strategy Governance Team consisting of:

- IT Director (sponsor with IT budget, technology expert)
- Deputy General Manager (original sponsor with program budget, business expert)
- Deputy Program Manager (solution and COTS knowledge)
- Program Manager (methods expert, facilitator, coach)

The Scope and Purpose of the meeting was:

The core-banking program wants to establish the best possible foundation for Workgroup management, where each Workgroup is responsible for establishment of its business "products" and "processes" to be supported by the COTS systems.

The non-traditional PQA workshop results were:

- An outline of the line organization managers with their areas of responsibility required for allocation of resources to the project organization:
  - Manager Fund Administration (Compliance, Settlement)
  - Manager Fund Management (Execution, Performance, Pools)
  - Manager Investment (Execution, Special Products, Trading Desk)
  - Manager Finance (Accounting, Management Information, Reporting)
  - Manager Treasury (Profit-Center, Bank Holdings)
  - Manager Account Management (Client Reporting, Fee Structure)
  - Manager Control Department (Client Static Data, Legal Reporting, Nostro Reconciliation, Treasury Control, Risk Limit Control, Counter Party Control, Lombard)
  - Manager Operations (Back Office Settlement/Reconciliation, Corporate Actions, Custodian Management, Broker Management, Securities Static Data)
  - Manager IT (Production, Infrastructure, Projects, Solution Support)
- The project organization (to be completed after the Activity Description workshop) will immediately commence the establishment of a complete acceptance test environment. The initial Workgroup work will be simulated Accept-Testing of real life business workflows produced by the Workgroups. The involved Workgroup participants will be asked to produce complete test cases before any future workshops.
- We immediately change the scope of workshops to produce only specific bank products with the quality required by the bank. In other words, we will change the scope from COTS functionality to bank business functionality.
- Each Workgroup will have precisely defined result objectives (agreed with the bank management as listed previously) to be obtained within an agreed deadline. These objectives will make it possible for each Workgroup to set up working sessions and to produce the results whenever this does not disturb the necessary daily work. There is no doubt that the highly qualified and experienced staff required in

the Workgroups will have to prioritize their time and make sure to deliver the Workgroup results on time.

- The Workgroup objectives will ensure a complete and feasible Information System solution, where the COTS contribution is precisely scoped and defined.
- Each Workgroup has the capability to:
  - Make necessary decisions
  - Initiate required work
  - Perform the work
  - Evaluate the work done
- This does not mean that all persons (roles) have to be available all the time, but it means that all roles and lines of communication are precisely defined beforehand. This will avoid any waste of time waiting for decisions to be made or waiting for facilities to be available.
- Workgroups will cover both business functionality and the IT environment. Assigned resources will typically work in more than one Workgroup, which will contribute to facilitating the communication between Workgroups.

3.12.2 The Non-Standard PQA Activity Description Meeting

The meeting with the departmental managers to establish Workgroups was called by a simple e-mail asking the following questions to be thought about before and during the meeting:

Who are the future users of the solution?
How should safe custody functions be handled and implemented?
What do you expect from the user interface?
How do we prevent the common reasons for IT failures?
Who can benefit from participation in project work?
How do you envision the implementation process?
How should availability, reliability, maintainability, and ability to integrate be ensured?
What methods, techniques, and tools should be available?
How should standard documentation be produced?
How can we ensure compliance with standards?
What standards do we have to develop or implement?
What are the education, training, and coaching requirements?

What is the biggest challenge concerning the future business operation?

What is the most common error and recovery problems encountered in the current business environment and what are their prevention requirements?

The very simple document standard to be used was attached to the e-mail with an example.

The General Manager opened the "PQA" meeting, but did not participate otherwise.

The Activity Descriptions developed in the meeting consisted of a PowerPoint presentation with all Workgroups and the business processes they were responsible to document in the form of the agreed Workflow-based standard documentation.

The document standard for the Workflow documentation was presented and explained to the participating departmental managers in the meeting.

The Deputy Program Manager and the Program Manager were declared Facilitators and Coaches for the Workflow writing process. We would support the writers, review, and approve the documentation for inclusion into the requirements specification to be delivered to the sub-contractors. The sub-contractors would be responsible for implementing the pertinent COTS functionality for the bank.

On top of this support, IT was committed to set up a sandbox environment with access to a test environment for all Workgroup participants.

The IT information System test environment was a dump on a specific date of all pertinent database content that could be accessed from all current systems in test mode. It was already possible to execute all business transactions inclusive of usage of external relations such as SWIFT and FIX protocols in test mode.

A similar test environment would be set up for the future COTS systems once these systems had been installed and parameterized for bank usage.

In this way, it was ensured that the Workflow documentation could show real life business processes with realistic screens and reports from the old systems. This contributed to high quality of the requirements specifications.

The Workgroups decided on in the meeting looked quite different from the ones envisioned by the Strategy Governance Team initially. This fact shows the importance of communication and involvement of the most competent persons when building teams and doing PQA. The Strategy Governance Team had done great work to get the most competent managers together to make the Workgroup setup decisions.

The Workgroups became:

1. Establish and Terminate Clients
2. Security Transactions
3. Cash Transactions
4. Derivatives Transactions
5. Corporate Actions
6. Tax
7. Fees (Income and Cost)
8. Control and Risk Management
9. Client Reporting
10. Infrastructure
11. Financial Control

The PowerPoint Activity Description for a Workgroup is shown in Figure 3.15 (Workgroup 2 has been used for this example; the numbers are telephone extensions).

The close out of the meeting was an agreement with all Workgroup managers to meet in the War Room every morning at 9:10 to discuss progress and constraints until all requirements had been produced.

All Workgroups had an end of May deadline, but we all knew that this was more than optimistic. Nonetheless, work progressed with an impressive speed and only minimal delays that were always reported well in advance of the planned delivery dates.

---

(2) Security Transaction Work Group

• Participants	• Scope
• NV (Back 501) WG manager)	• Order, Deal
• HS (Trade 485)	• Securities Static Data
• EF (Back 580)	• Prices
• KO (Fund M 336)	• Booking
• PJ (Front 231)	• Mortgage
• SS (Finance 362)	• Confirmation, Reconciliation,
• LJ (Cash 6221)	Settlement
	• etc.

FIGURE 3.15
Bank case Workgroup Activity Description.

The minutes from the first 9:10 meeting and the call for the second 9:10 meeting were:

Mail or call me if your Workgroup will not be attending the next 9:10 meeting, please.

We will discuss important experience from work in Workgroups 2, 1, and 7, who have initiated their work. Any other work experience is also presented and discussed.

The 9:10 meeting agenda:

1. Workgroup progress
2. Workgroup issues
3. Workgroup coordination
4. Documentation standard implementation

Please look at I:\xxxxx with room for your progress.

Please take a note of the time you use for defining and documenting test cases. A system will soon be in place for automation of this registration, where we need the time already used now to be registered, please.

Workgroups have now been decided and already slightly changed as shown above:

- XX from Back Office is now member of 9 covering Client Reporting
- YY from Finance is now member of 11 Financial Control

It is a key to our success in the program that all Workgroup Managers participate in called for 9:10 meetings, as these meetings in the early stage are used for ensuring that the Workgroups get started off in the right direction and approach their targets as fast as possible.

Because some project work has already been done in earlier workshops with focus on COTS functionality, it is now difficult to change this established culture—but we have to do that in order to reach our target of having the Core Banking Solution in production by November.

The current focus is BUSINESS PROCESSES END-TO-END.

This means that we do not try out business processes on the future core-banking information system solution until the COTS systems have been installed and adapted to these business processes.

The user Workgroups define business processes, test data, and business workflows to be tested in the future COTS-based information system environment:

- Business processes are described using the delivered documentation standard—quality to be established in practice.

- Test data are presented in screen dumps from the current information systems and the test cases representing all relevant variation are listed in Excel tables.
- Expected output can be a listing, a file dump, or a screen dump with reference to the test case giving this result.

Once a set of test cases representing a business process end-to-end has been created end approved by the complete Workgroup and your coach, it is delivered to the IT Workgroup.

The business process Workflow must, wherever possible, document required improvements to the current established business processes, especially where this concerns contractual or legal compliance.

The IT Workgroup (yet to be established) sets up COTS and other involved software to simulate the business process before the users are allowed to test the business case in the future environment. We will not waste the users' time with endless demonstrations of COTS functionality, which they do not need. We only show what works for the bank users in real life.

If the COTS applications should not be able to improve the users' current working conditions or solve the tasks required from the workflows, the problems are discussed in the 9:10 meetings and documented in an outstanding list for problem and change management.

In this context, we have already seen problems with the possibility to set up screen images corresponding to users' needs and with reuse of codes and parameters from the current IT application environment, which indicates a potentially difficult future usage and learning process—not even mentioning conversion and integration problems.

See you 9:10!

Please send a deputy if you are not available. We will ensure that all Workgroup managers get together every morning just for a few minutes to discuss the progress and eventual issues.

3.12.3 Complications from Not Following Standard PQA

While the non-standard PQA result as such proved very efficient for the success of the Strategic Initiative implementation, it did not prevent all problems during the implementation of the COTS systems and the information systems using the COTS systems.

The manager of Finance was new in the organization and did not participate in the initial PQA Workgroup setup. Finance is more important in a banking environment than one might expect. All banking transactions are reflected in some financial transaction concerned with, for example, fees from clients and funds, where the bank is the selling agent, and with, for example, clearing costs from custodians and stock exchange agent services, and much more. Besides this, all human resources related costs including relationship management commission and normal business investments in IT and administration are handled by Finance.

Finance fully understood its central role during implementation and testing, but it caused many delays based on missing competence in the organization (the manager did not have banking competence).

The problems could have been avoided if Finance had made its role clear initially, but it refused to do that and demanded to implement the financial side of the solution on its own. It did not use the core-banking system for finance, only to feed the financial control COTS system that was procured as an additional requirement to the program scope.

In the mind of the Finance department, the implementation of the core-banking COTS system was responsible for explaining to it what transactions were fed to Finance without any responsibility on the Finance side.

The Finance arguments were strong and general management approved its requirement to be autonomous.

During the Accept-Testing situations, the contribution from Finance proved that this autonomy was a bad decision. Finance was never ready for testing because it did not have the time to understand the full financial implications of the bank transactions. Finance caused program delays of several months.

The bank risk manager was not involved in the initial PQA because he was occupied elsewhere. Unfortunately, we only learned about this ten minutes before the meeting started, so we could not send all other participants home. The risk manager refused to cooperate with program management or with any Workgroup in the program.

In the initial non-standard PQA meeting, the bank risk manager was appointed Workgroup leader because nobody else understood this very complex responsibility.

Risk Management expected to get all their required reporting from IT without presenting requirements other than copies of their current reports. The why and even the what and the how of these reports were not documented, so even the best programmer and report designer in IT could do nothing. On top of this, the old reports were completely outdated from a banking risk management standard view (Basel rules). General management again approved this situation and just expected IT to deliver.

For this reason, Risk Management was established as a Workgroup, but it never produced anything.

In order to improve on this situation, the program manager wrote an appropriate Activity Description for the Risk Manager, who did not like this at all.

SCOPE

Risk management is a core activity of banking. In the context of Core Banking, the Risk Management solution is implemented comprising:

- A documented and management signed off Risk Management Policy
- A set of documented and approved Risk Management procedures, which visibly comply with and fulfill the Risk Management Policy (the Bank Risk Management User Guide to be established)
- A set of documented tools in support of the Risk Management procedures (reports in and outside COTS, usage of and compliance with external and corporate information systems, etc.)

At a minimum, the currently implemented and documented risk policy and procedures must be reviewed and brought up to current Best Practice, for example, BASEL principles/standards (Basel states: "Clear strategies and oversight by the board of directors and senior management, a strong internal control culture [including, among other things, clear lines of responsibility and segregation of duties], effective internal reporting, and contingency planning are all crucial elements of an effective operational risk management framework for banks of any size and scope").

DELIVERABLE

- An implemented agreed to Risk Management Policy for the bank that ensures organizational anchoring and ongoing improvement in accordance with best practice.
- Implemented agreed to Risk Management procedures.
- Tested, agreed, documented, and implemented COTS-based tools in support of the agreed to Risk Management procedures. This comprises the availability of a Bank Risk Management User Guide and Training with training material.

PURPOSE

The implemented Risk Management solution will provide the bank with documented best practice concerning the specific bank requirements for Operational Risk Management.

3.12.4 What We Could Have Done Better

One important success factor concerned with Activity Description preparation is that the Activity Leader (Sponsor) takes ownership of this description and the activity. In this way, the motivation of the sponsor is ensured and the people working to deliver the activity products get the best possible support in the form of knowledge and funding.

The lesson learned here is that only Activity Descriptions prepared by the persons directly responsible for getting the activity executed will be reliable, not the ones prepared by the Program Manager.

What I did here took away all opportunities to establish a feeling of ownership from the primary responsible Activity Manager, which I know very well is the key to success with any activity.

The program did close out successfully with serious delay because competent resources to close the holes in the organization were added later.

If I could redo this program, I would add a Program Office with competent resources to ensure appropriate requirements where the bank organization cannot provide these on their own. The Program Office could have the following resources:

- Business Analyst
- Banking Risk Expert
- Compliance Expert
- NAV Calculation Expert
- Finance and Tax Expert
- General Banking Operation Expert
- Reporting and Communication Expert also covering Program Performance Reporting

The Strategy Governance Team is a decision-making organization. They learn from Program Office findings and they have access to pertinent progress information prepared by the Program Office on the basis of tracking information from Workgroup managers and the Project Office. The Program Office works full time to establish pertinent requirements and to track specific program or Strategic Initiative progress.

The lack of a Program Office resulted in a too heavy work burden on the Program Manager and even on the Deputy Program manager and the other members of the Strategy Governance Team.

3.13 THE MILITARY HEALTHCARE INFORMATION SYSTEM CASE

The establishment of a Military Healthcare Management system is—as you would expect from a military-based case study—a PQA example of how PQA is done most efficiently.

We became responsible for conducting this project because we were well established in the Defense Facility Management, where we had implemented the common information system and produced all the handbooks and training material for the education and certification of Defense Facility managers.

The key-stakeholders here were not the facility managers, but the defense high command represented by doctors on a colonel level in cooperation with an army colonel responsible for the healthcare service information system implementation. The army colonel was our sponsor.

A group of military doctors and dentists (all of them military officers on a high level), the sponsor and his deputy, and a business analyst from my organization formed the Strategy Governance Team.

The sponsor and I developed the PQA Introduction for the Strategy Governance Team.

The Strategy Governance Team visualized and documented the complete project scope and established the Workgroups that developed the requirements specification.

As there was no prior experience with a common military healthcare information system, the requirements specification turned out to be a fully designed healthcare information system solution ready for programming or for implementation in a COTS application, if such one could be found.

PQA as a method was fully approved by all stakeholders for planning and project management.

There was absolutely no stress to provoke deviations from standards, and the result became excellent.

This case, just like the banking case, will also be used to show you the execution of business analysis and the preparation of core elements in the solution requirements spec, which will be shown in Chapter 5. Beside PQA facilitation, we also took on the role of business analyst and solution designer in this case.

3.13.1 Strategy Governance Team PQA Introduction

The Strategy Governance Team is also the team of people conducting the PQA Workshop.

The PQA Workshop Introduction (without the standard PQA process presentation) looked like this:

Implementation of an Information System for Defense Healthcare (DHS)

PQA INTRODUCTION (EXTRACT)

2.   SCOPE AND INTRODUCTION

2.1 DHS AND THEIR CURRENT INFORMATION SYSTEMS

DHS core tasks are:

- Prevention and treatment of diseases in and injuries to military personnel. The preventive work is primarily focused on sports and work medicine. The treatment of diseases and injuries are primarily handled in the place of work infirmaries and dental clinics. In peacetime, the patient will be referred to the civilian healthcare system if the patient has a need for more complex treatments.
- To set up disaster and war preparedness, which during crises and war conditions allow the military to convene a large number of doctors and other healthcare professionals who have been trained in the military, and to establish field hospitals.
- Training of healthcare personnel, which takes place in schools and training units.
- Support to civilian authorities in the form of helicopter and sea rescue services, design of life-saving equipment at sea, etc.

DHS employs 125 people and has an annual budget of $45 million. To this must be added the permanent healthcare personnel at places of work consisting of approximately 300 physicians, nurses, dentists, clinic assistants, medical officers, medics, and others.

There are more than 50 infirmaries in defense distributed on the Army Operational Command, the Tactical Air Command, and the Navy Operational Command with operational ships.

DHS provides annually more than 200,000 medical consultations and more than 142,000 dental services.

In some places of work, it is possible to implement small-scale research projects in the form of drug testing, sports medicine studies, etc. DHS does not have a dedicated research program. The research projects are conducted as needed. The DHS annual budget for research amounts to approximately $10,000.

To support the execution of DHS tasks, the following healthcare information systems are currently used:

FMJS, the Pilot and Diver Medical Journal System

WONCA, which registers all infirmary treatment with a WHO diagnosis code

FLOTRE, the Armed Forces Medical and Dental Healthcare Personnel register is not yet in use.

FMJS is today used solely in infirmaries under the Tactical Air Command. It only records the health status of the pilots even if the system is also prepared to be used for recording of the health status of divers.

FMJS manages healthcare information on all pilots. The main functions of FMJS are appointment-booking control for periodic medical examinations and administration of medical records. There are approximately 25 users of FMJS.

From a purely practical point of view, FMJS meets the information and management needs that exist in the area, but the technical platform for FMJS is quite outdated with a high derived risk of malfunction. The maintenance contract of the system has been terminated.

WONCA is used for recording of all inquiries to infirmaries in the form of a diagnosis (WONCA code). The used WONCA codes represent approximately 400 of the most common diagnoses out of the approximately 30,000 WHO diagnoses used in the civilian healthcare system.

Besides the WONCA code the patient's name, social security number, and place of work are also recorded. These data are used for periodic reporting to DHS and external public healthcare organizations.

Both DHS and the places of work use data from WONCA in their establishment of objectives and activity frameworks.

At DHS, there is a presumption that the number of inquiries to the infirmaries is greater than the actual number of records in WONCA, which can be attributed to the current registration routines in the infirmaries.

Although it is now possible to register performed dental work in WONCA, this facility is not used in the dental clinics. The dental clinics use paper and pencil for their records and reporting.

FLOTRE is intended to be used to record more detailed data about the doctors', dentists', and nurses' healthcare professional background (medical specialties, etc.). With this information recorded, it will be possible to account for the personnel's current level of competence in connection with, for example, work placement under mobilization.

Master data in FLOTRE will be updated annually based on completed questionnaires from individual physicians, dentists, and nurses.

FLOTRE is currently not operational.

Apart from the above-mentioned information systems, locally purchased civilian electronic medical record systems are used in some infirmaries and dental clinics, which reflects the latent demand for such information systems. The problem with these COTS applications is that they most often rely on proprietary databases that in most cases do not provide possibilities for exchanging data with other external and internal systems.

The Workgroup on Medical Programming (WG Medpro) has produced a "Report on Electronic Health Recording Systems at DHS." It confirms that the IT-based information systems used by DHS and the place of work infirmaries and dental clinics have not been implemented because of a predetermined data processing strategy, but rather ad hoc as the needs arose.

This has resulted in that the IT-based systems used are isolated "data islands" with no or very little value to the work processes and information needs within military healthcare.

2.2 WHY DHS NEEDS AN IT-BASED INFORMATION SYSTEM

DHS has called for the implementation of a common military healthcare information system (FOSIS) to replace all existing healthcare information systems that are used in places of work, thereby reducing the number of "manual" systems (especially the paper-based patient journals) and improving the quality of DHS services in general.

The primary objective is to implement an EPJ or Electronic Patient Journal. The EPJ must be standardized with a view to international usage to handle all patient data including ECG and X-ray images with a maximum of security.

The EPJ must be able to support the entire process associated with the treatment of patients in infirmaries and dental clinics, also in the context of mobilization (e.g., military field hospitals).

The EPJ must be able to operate geographically independent of the place of medical treatment so that patient data is not lost when such personnel are transferred to other places of work. The patient must have the opportunity to be treated in any infirmary or dental clinic with subsequent direct update of the patient's EPJ.

2.3 WHAT WE WILL ACHIEVE IN THE PQA WORKSHOP

The purpose of this PQA Workshop in the DHS Strategy Governance team is to establish the project team and to define the framework for the implementation of FOSIS. The participants designated to form the Strategy Governance Team will define as detailed as possible why, what, and how concerning their requirements to FOSIS:

- The work situation and the results that they want to achieve with FOSIS
- The process that the project participants and future project participants must follow in order to implement FOSIS
- Current and future project participants' involvement, motivation, skills, and competences

We will not formulate specific technical requirements to the future solution as these requirements will be precisely defined by competent resources task by task later on.

However, we will define and visualize the targets for these resources to be able to work effectively and produce efficient solutions, which can and will be accepted by DHS management and in the infirmaries and dental clinics. Relevant ideas and suggestions are however welcome.

2.4 HOW YOU PREPARE YOURSELF FOR THIS WORKSHOP

Your participation on the PQA Workshop allows you to express and demonstrate how you think the FOSIS project success can be ensured. You give inspiration to and get inspiration from the other PQA participants.

You must express:

- Your personal vision of the future FOSIS in DHS
- Your interpretation of the mission of the FOSIS Project Governance Team

Based on your experience with and knowledge about your defense environment, please prepare an expression of what you consider the most extreme case of success for the procurement and implementation of FOSIS. Describe the factors especially relevant to your responsibilities and competences in your organization and within the FOSIS Project Governance Team. Please be prepared to explain your points of view.

2.5 AREAS OF CONCERN

The following areas of concern are examples of an inspiration to the subjects that we will consider under the workshop. Please remember to express why you intend to answer the question the way you do:

Who are the DHS clients?

What do the clients expect from DHS?

Can a common system solve the tasks for medical treatment and dental treatment?

What demands from the army are not relevant for navy and air force?

What demands from the navy are not relevant for army and air force?

What demands from the air force are not relevant for army and navy?

What are the areas of work where DHS can be more productive?

Who must participate in the definition of the detailed requirements to FOSIS?

Can DHS and the defense organization provide the competence required to succeed with FOSIS?

What external organizations and administrations must be involved during implementation?

Will we improve the administrative procedures in and under DHS before we implement FOSIS, or in parallel with implementing FOSIS?

Will DHS be an international organization under The Danish International Brigade?

What are the demands to the IT infrastructure in support of FOSIS?

What DHS procedures must be supported by FOSIS?

Will DHS be certified ISO9000?

How can FOSIS improve the DHS Quality Management?

We look forward to a rewarding workshop.
Kind regards
LI1 and LI2

3.13.2 Strategy Governance Team PQA Workshop Result

The PQA Workshop result looks like this:

DHS Implementation of
FOSIS
for
Defense Healthcare
Project Quality Assurance

Result
January

Participants

Dent	MW	DHS (Project Manager)
Mil	AF	DHS
Dent	MH	DHS
Doc	OC	DHS
Doc	KT	DHS
Mil	IB	D IS (Sponsor)
Off	EH	D ID
Mil	TN	D IT

Facilitation

SL	LI
CH	LI

1   WHY DHS NEEDS AN IT-BASED INFORMATION SYSTEM

DHS has called for the implementation of a common military healthcare information system (FOSIS) to replace all existing healthcare information systems that are used in places of work, thereby reducing the number of "manual" systems (especially the paper-based patient journals) and improving the quality of DHS services in general.

The primary objective is to implement an EPJ or Electronic Patient Journal. The EPJ must be standardized with a view to international usage to handle all patient data including ECG and X-ray images with a maximum of security. The EPJ must be able to support the entire process associated with the treatment of patients in infirmaries and dental clinics in the context of mobilization (e.g., military hospitals).

The EPJ must be able to operate geographically independent of the place of medical treatment so that patient data is not lost when such personnel are transferred to another place of work. The patient must have the opportunity to be treated in any infirmary or dental clinic with subsequent direct update of the patient's EPJ.

2 INDIVIDUAL VISIONS

OC

A system for the entire military healthcare, which:

- Meets the documentation requirements for examinations and treatments.
- Facilitates and streamlines workflows locally as well as exchange of information with the civilian healthcare system.
- Ensures access to information for relevant persons at all levels, thereby providing complete and reliable basis for the ongoing healthcare advice.
- Ensures opportunity for statistical processing, etc. for use by the healthcare quality assurance and research locally as well as in the defense as a whole.

MH

FOSIS should be simple, easy to use, and be compliant with legislation for doctor' and dentists' work.

FOSIS data must be stored and secured, and thus at all times be available to relevant persons search on data.

FOSIS must be immediately compatible with civilian healthcare systems with flexible communications in mind.

Moreover, FOSIS must provide support in the daily work, including work organization, personnel management, and booking of patients.

FOSIS data must be recorded continuously, so that the management has updated information about:

- Production
- Consumption
- Waiting times

All of this is desired in order to be able to calculate the productivity/efficiency and capacity utilization.

KT

A single, approved registration system for all military healthcare services that meet any registration requirement:

- Manageable, clear, systematic sanitary professional documentation
- An effective management tool in the general and local planning
- Instrument to extract data in any ad hoc relationship (research, epidemiological and demographic data)
- Systematic exchange of information between management and users at all levels
- Quality assurance of healthcare services and of the healthcare organization

- Can communicate in the context of current and future communication facilities (fixed line and wireless communications) under both military and civilian conditions

MW

FOSIS must be a single system for all healthcare personnel in the Armed Forces, which is based on standards from the civilian healthcare sector, facilitating cooperation in the individual infirmaries and between the Forces. FOSIS should be able to solve the tasks everywhere.

FOSIS must meet the information needs with respect to the following:

- Daily patient care
- Electronic medical bag
- Decision support (Knowledge Couplers)
- Access to knowledge databases
- Operation support
- Patient Administration
- Planning
- Communication (internal/external)
- Management Information System
- Staff management
- Education
- Use
- Resource allocation
- Production
- Time usage
- Services
- Quality
- Data for research

AF

FOSIS must be a future-proof system (electronic system) to be able to accommodate all current healthcare systems in defense.

IB

1. Architecture
    - Common system
    - Operate on network and standalone
    - Widely usable COTS

2. System Application
    - Electronic medical bag
    - Electronic journal
    - Data extraction
    - Update
    - Support remote diagnostics
    - DB
    - Registration

- Reporting
- Statistics
- Diagnosis
- Replace existing systems (FMJS)

3. Management Information (MIS)
   - Generate relevant KPI
   - Support answering of questions from the political level

4. Communication
   - Military and civilian networks
   - In addition to alphanumeric data also imaging and X-ray image transmission

5. General
   - Confidential P
   - Used in garrison and under field conditions

EH

A user-friendly, mobile, and centralized paperless journal system appropriate to the defense IT strategy, including the integration with the De Mars management system:

- Intelligent study support
- The system must comply with all laws and safety requirements
- Exchanging information with external systems

TN

FOSIS must be a COTS product that is easy to work with, which meets DHS requirements, which can cooperate with other relevant defense systems, including the De Mars system, which must be well documented and easy to learn for users.

3. CRITICAL SUCCESS FACTORS AND THEIR SUGGESTED SUCCESS FACTORS

1. Essential FOSIS functionality implemented simultaneously on time.

1. The system is implemented simultaneously by all users.
2. The suggested solution must be accepted by DHS management before March 1.
5. Interdependent system components are implemented simultaneously.
10. The budget for FOSIS must be approved before March 1.
18. The tender must be completed by July 1.
25. The contract must be signed October.
32. The first phase of FOSIS must be implemented by the end of December.
47. FOSIS must able to be implemented in steps.
50. The annual maintenance fee for FOSIS may not exceed 15% of the purchase price.
54. Usage of FOSIS must be measured relative to well-defined milestones.

55. The future users of FOSIS must not lose their currently recorded electronic data.

2. Intuitive Danish language user interface.

3. Easy to use (user-friendly).
11. Fast to learn user functions.
12. The system should be easier, faster, and smarter to use than a paper-based system.
16. FOSIS must for each information user be able to present the necessary information in an understandable and clear manner.
57. FOSIS user interface must be in Danish.

3. FOSIS supports all healthcare services throughout.

4. The system must be able to support the healthcare services throughout.
21. Mobility without data reduction.
23. Procedures for full use of historical information.
26. FOSIS is able to provide operational support to:

- Patient Administration
- Planning and Resource Management
- Management of healthcare personnel

36. FOSIS must able to support DHS as well as medical and dental service in peace and the war structure under crisis and war (advice and workflows).
43. FOSIS must support the clinical decision process.
51. FOSIS must be a system common to medical and dental service in the Armed Forces.
52. FOSIS can be used down to echelon 2 level.
53. FOSIS must support centrally managed healthcare information, be it defined a priori or ad hoc.
58. FOSIS must replace all existing healthcare IT systems in defense.
59. FOSIS must be able to handle relevant healthcare management information.
61. FOSIS shall for each information user be able to present the necessary information at the right time and place.

4. FOSIS provides access to necessary and complete healthcare information.

7. The system must always be updated and contain all valid healthcare information.
8. Consistent data.
31. FOSIS must be able to generate the necessary product reporting key figures.
33. FOSIS must be able to generate disease-and performance patterns for DHS.
37. FOSIS must eliminate the loss of all types of healthcare data.
41. FOSIS must ensure the validity and as complete data as possible.
55. The future users of FOSIS must not lose their currently recorded electronic data.

5. FOSIS communicates with relevant systems.

14. Well-defined interface to De Mars.
17. FOSIS must be able to communicate with civilian healthcare systems.
19. The system must be able to communicate among peers and with senior authorities.
24. FOSIS must be able to exchange data with the Conscription Agency's IT-systems.
35. Data must be retrievable to be exported to analysis tools.
38. FOSIS must be able to receive data from the health cards, citizen cards, etc.

6. FOSIS meets the defense IT strategy and is based on relevant standards.

15. A flexible system that can be quickly adapted to changing requirements.
20. FOSIS must be based on civilian standards, such as SKS, EHCRA, HEP, and MEDCOM.
22. FOSIS must comply with the defense IT strategy.
29. FOSIS must be COTS.
40. Using standardized rules for journal record completion and reporting, FOSIS facilitates the access to epidemiological and demographic studies and research.
45. FOSIS must be a common system (uses the common technological infrastructure of the defense).

7. FOSIS meets all requirements for safety and traceability

9. Access to the data stratified.
30. There must be automatic logging of who has used the data when.
39. FOSIS must be acceptable safety wise by defense intelligence.
44. Full traceability of data in patient journal records and diagnoses.
62. FOSIS must for each information user be able to present the necessary information with high data quality and data security.

8. FOSIS enables a flexible, user-specific data handling.

23. Procedures for full use of historical information.
28. Clear longitudinal data presentation.
34. Optional chronological or problem-oriented journal.
40. Using standardized rules for journal record completion and reporting, FOSIS facilitates the access to epidemiological and demographic studies and research.
42. FOSIS must be able to handle hierarchical and self-selected workflows.
48. FOSIS must support a convenient, detailed, quantified activity registration.
49. FOSIS must be able to establish ad hoc defined arbitrary groups and action plans.
60. FOSIS is able to handle ad hoc definition and handling of data.

9. FOSIS increases quality and efficiency in healthcare service

6. The system must be paperless.
13. All healthcare service data must be contained in an electronic medium.
27. The introduction of FOSIS must be able to free up resources for solving of pt. not solved healthcare tasks.
46. FOSIS must serve to enhance the credibility of the defense medical and dental services for patients, Armed Forces top management, and civilian healthcare bodies.
56. User responsibilities and organizational structure supports an efficient use of FOSIS.
63. FOSIS must for each information user be able to present the information necessary at minimal cost relative to price and usage of time.

4. PQA MATRIX

CRITICAL SUCCESS FACTORS

9	FOSIS increases quality and efficiency in healthcare service
8	FOSIS enables a flexible, user-specific data handling
7	FOSIS meets all requirements for safety and traceability
6	FOSIS is aligned with the defense IT strategy and is based on relevant standards
5	FOSIS communicates with relevant systems
4	FOSIS provides access to necessary and complete healthcare information
3	FOSIS supports all healthcare services throughout
2	Intuitive Danish language user interface
1	Essential FOSIS functionality implemented simultaneously on time

Activities		1	2	3	4	5	6	7	8	9	Value	Responsible
1.	Provide decision basis	*									3	EH
2.	Provide needs assessment			*	*	*			*	*	2	EH
3.	Prepare requirements specification		*	*	*	*	*		*		2	TN
4.	Prepare System Specific Requirement Statement (SSRS)							*			0	TN
5.	Prepare tender documents	*	*	*	*	*	*	*	*	*	0	TN
6.	Complete procurement	*									0	TN
7.	Establish project organization	*		*							0	EH
8.	Manage the project	*									0	EH
9.	Implement phase I	*	*	*	*	*	*	*	*	*	0	TN
10.	Inform stakeholders	*		*						*	1	MW

3.13.3 Activity Description for FOSIS Implementation

These two Activity Descriptions were approved by the PQA Team to be executed immediately in support of procurement of the future COTS system that the team hoped to find. We were deeply involved with the Workgroups that performed the activities. As mentioned previously we performed business analysis and solution design to be used directly for tendering.

The tendering process would follow European Union standard rules. We were not at all allowed to be involved with the actual tender because we had been too involved with the production of the tender material production.

Here are the two Activity Descriptions:

Activity Description 3	Prepare Requirements Specification
By: TN	Approved by WG: January
Scope	Prepare a requirements specification (RS) for the Military Health Information System (FOSIS), which based on Project Quality Assurance (PQA) and Information Requirements Study (IRS) must describe all the requirements that users have to FOSIS. This RS is the basis for a future procurement. The main features of the IRS are: • Definition of functional areas • Definition of functional areas objects • Efficiency criteria • Documentation of communication needs • Requirements for new procedures and systems
Products	The RS, which describes the overall functional requirements for use of FOSIS at DHS and other authorities. Requirements for distribution of responsibilities and task solving in the organization to use FOSIS are documented in order to ensure optimal use of FOSIS. Furthermore, a cost/benefit analysis for FOSIS is documented. Finally, the overall requirements for FOSIS structure and functionality are documented with a suggestion of a phased implementation of the system.
Purpose	The preparation of the RS shall ensure that all stakeholders' information needs in the performance of healthcare services are identified and documented, so these information needs can be related to the desired functionality in FOSIS to support the handling of the information needs.
Responsible	DIT
Other resources	DHS employees, WG FOSIS participants, DIS and DIT representatives, LI

Sub-activities	Task Description	Resource Needs
	1 Clarify the analysis structure and define functional areas.	DHS, WG, LI: 1 day
	2 Carry out analysis of necessary functional areas	DHS, WG, LI: 40 MH per functional area
	3 Review of section reports on management level	DHS, WG, LI: 24 MH per review
	4 Preparation of the requirements specifications	DHS, WG, LI: 120 MH
Time frame	March to May Incl.	
Risks	It can be difficult to release AG, DHS, NIV, the resources for this work, which could delay implementation. The necessary financial resources of Defense High Command are not allocated. Because of the desired FOSIS functionality, Defense Intelligence cannot approve further work.	
Dependency on other activities	Predecessors: 2: Provide needs assessment Successors: 4: Prepare SSRS 5: Prepare tender documents	

Activity Description 5	Prepare Tender Documents	
By: TN	Approved by WG: January	
Scope	Prepare tender documents for FOSIS based on PQA, IRS, and RS. This must be taken into consideration for a supplier's offer. The tender documents shall describe the requirements for a COTS product that takes DHS requirements in consideration.	
Products	Tender documents with built-in ready to use contract "only" to be signed.	
Purpose	To get a number of suppliers to offer the FOSIS solution based on the tender documents by delivering a COTS product, manage the implementation according to the implementation plan, and be responsible for the training of DHS and other authorities' personnel in optimal use of the solution.	
Responsible	DIT	
Other resources	DHS employees, WG FOSIS participants, DIS and DIT representatives, LI, COTS vendors	

Sub-activities	Task Description	Resource Needs
	1 Logistics/schedule	WG, DIS: 8 hours
	2 Legal conditions	DIT, DIS: 4 hours
	3 Requirements to response content	DIT, DIS: 4 hours
	4 List of vendors to be advised directly	DIT, DIS, WG: 16 hours
	5 Demand to vendor quality system	DIT: 4 timer
	6 Ready for use contract	DIT: 8 timer
Time frame	June	
Risks	It might be difficult to release WG, DHS, and Level I resources for this work, which could delay implementation. The necessary financial resources for the procurement of the COTS solution are not allocated, which might stall the project. Because of the desired FOSIS functionality, Defense Intelligence cannot approve further work.	
Dependency on other activities	Predecessors: 3: Prepare requirements specification 4: Prepare SSRS Successors: 6: Complete procurement	

3.14 LESSONS LEARNED

Risk Management performed efficiently can allow the teams involved with Strategic Initiatives to build plans that with higher probability achieve the solutions and results (the impact) demanded by the stakeholders.

We are constantly faced with pertinent unknown unknowns and unknown knowns that are ready to surface at any point in time in the future of our Strategic Initiative.

Risk responses are built into the project plan as improved activities or new activities to avoid the risk or to mitigate the negative effect of risk.

The Strategy Governance Team establishes and maintains the master plan that binds together and integrates all the subsequent plans. The Strategy Governance Team establishes the master plan as the top level PQA Team. In this top level PQA process, they establish the first level of subordinate PQA Teams that together ensure the delivery of the required strategically aligned solution components.

Active involvement of all pertinent and available knowledge from inside and outside of the organization in the PQA process ensures agreement to the established objectives for the Strategic Initiative.

You need supplementary procedures and tools to PQA if you want to know where you are compared to where you want to be while executing a Strategic Initiative.

The needs of corporate stakeholders are very different, and most often inconsistent and contradictory.

PQA and other SQM processes such as outlined previously ensure completeness and sufficiency of the solution, the process, and the organization required to achieve the benefits expected from implementation of the strategy.

A Success Factor expresses a pertinent wanted and needed attribute of one or more quality objects, while a Critical Success Factor is a class of Success Factors.

The Strategy Governance Team knows and understands the conditions and the needs of the corporation in its current situation based on a thorough SWOT analysis or based on the mutual knowledge and experience of the team members.

PQA visualizes the motivating factors of the key-stakeholders by committing them to document and present their personal vision of the result of and their personal view on their mission during the Strategic Initiative.

The common acceptance of initial objectives and the initial strategy of activities to achieve these objectives ensure an opportunity for a "no conflict" implementation process.

The PQA rules have been established in order to ensure active involvement of all participants. No one is accepted just as a guest or to listen in passively. On the contrary, the rules are there to ensure the synergy that is only possible if the explicit and tacit knowledge of all participants are provoked to be used.

It is one of the keys to successful conduction of a Strategic Initiative that the top-level stakeholders and the original sponsors are involved and activated in the initiative activity whenever this gives them an opportunity to show their personal motivation and support of the PQA Teams.

All participants in the initial PQA Workshop participate in the first PQA review in order to ensure that agreed activities are complete, valid, and consistent with regard to success of the Strategic Initiative.

Think twice before you bring in new resources in order to speed up things—most often new resources bring more chaos than progress. If you need additional resources, make sure to plan for them and to allow for them to be competent and to be adapted to your activity. This applies to people and technology.

The Workgroup Team assists in the production of the Activity Description to be reviewed by the PQA Team in order to ensure valid and reliable estimates and quality expectation.

If you do not have access to Project Office support, you are obliged to establish this support, even if you have to create your own Project Office.

The Strategic Initiative planning workshops cannot yield a usable result without a fully up to date project management system database that is supported by the Project Office in support of the meeting discussions and decision making.

Only Activity Descriptions prepared by the persons directly responsible for getting the activity executed will be reliable, not the ones prepared by the Program Manager or the Facilitator.

The lack of a Program Office resulted in a too heavy work burden on the Program Manager and even on the Deputy Program Manager and the other members of the Strategy Governance Team.

4

Solution Provider Procurement

I have seen quite a few enterprises dealing with resource procurement just as if they were doing recruitment of employees. This is acceptable in situations where it is difficult to find qualified people as employees or where you need to replace an employee for a shorter period. However, it is not a good idea if you need resources to produce a specific solution to a complicated task, such as the setup and implementation of a COTS-based information system.

The resources you need for setup and implementation of COTS-based information systems are highly specialized and qualified people who can contribute to your total solution with an important part of an information system foundation, but you do not need them every day once the solution is in production.

Take a simple example such as the procurement of a car:

You do not employ engineers and technicians that produce your car; you just buy the car.

You can buy whatever service is needed to make the car run from a car service provider.

You might employ a driver to run the car or you might drive it yourself.

In this example, you procure a solution. You do not procure engineers and technicians because you would never be able to lead and manage them to produce what you need. Furthermore, you would not be able to provide them with the production environment needed.

In car procurement, this is obvious; when it comes to COTS system implementation, it is less obvious, but just as true.

Many organizations think that buying a COTS system is like buying a car once they have a number of product user licenses and a system operator in place.

They believe that once the COTS system has been delivered and runs on their IT infrastructure with access securely ensured to the users, the users just turn on their PCs, laptops, or other connected devices and run their business processes.

Even the most simple accounting or project management COTS system does not allow you to just turn the key and run even after training and operation management setup.

In order to be able to use and manage IT- and COTS-based information system solutions, you need competent users and an IT organization with three basic components:

Information System Management	IT Infrastructure Management	IT Service Management
This is the COBIT (www.isaca.org/) structured environment that ensures that business processes and users are supported to run the corporate business in an optimal way using their available information systems. This environment manages, for example: • Requirements specification • Solution development • Solution implementation • Solution governance	This organization runs the IT infrastructure using ITIL principles. It can be internal or it can be outsourced partially or completely. Cloud usage management would be part of this environment. Daily COTS operation is handled here with, among other tasks, the consolidation and reconciliation processes that ensure integrity and validity of data across systems and periods: • Availability supervision • Performance supervision • Incident handling • Problem management	This is the ITIL (http://itlibrary.org/) structured environment that ensures that the IT infrastructure is acquired and supported to be available and performing as required by the users. IT and COTS vendors are managed here: • Service level agreements (SLAs) • Release management • Change management • Error handling

A COTS system is not your information system solution.

What you want to procure is not a COTS system, but a solution, which is a business information system running just like a car; once the rules of conduct are established, the users are trained and sometimes certified, and the technical operators are competent to run the solution. You can use the ITIL and COBIT structure of organizational elements, processes, and objectives as a checklist to verify that your business organization and your IT organization cover these elements.

The COBIT standard comprises an Acquire and Implement section. You can get inspiration from this section to ensure that you do the right things, but the standard will not tell you how to do the things right.

The objective here is to show you an example of a relatively complex procurement of COTS systems and Solution provider services that went well under specific conditions, not to explain to you how to procure under all circumstances.

Procurement establishes a future required situation, in our case:

- New COTS systems delivered and implemented
- Solution provider service level agreements
- User training
- System operation

In order to succeed with procurement, you need to manage the risks involved with this process. Some of the more important risks comprise:

- If you buy a COTS system before you have a detailed requirements specification, you will with high probability waste time and money because setup and implementation to fit your (unknown) needs will be a trial and error process until a feasible solution is established with very low probability of success.
- If none of the potentially available COTS systems can contribute to your information systems needs without major changes or additions to their functionality, you might get important cost increases and solution delivery delays. In this case, we do not talk about parameterization and setup changes, but about changes not supported by the delivered COTS functionality.
- A chosen COTS vendor always obtains a de facto monopoly once chosen and installed in your IT environment. You will be very weak in negotiations and might incur long lead times and high costs for adaptations in support of your business operations, especially facing business needs that are particular to your business if you do not foresee and include the conditions of these changes in the vendor contract.
- If the COTS vendor goes bankrupt or in other ways ceases to do business, you lose support of your COTS system and you might need to procure another one. This is the background for the escrow clause in the vendor contract that is explained next. The escrow does not

prevent you from losing money and time, but it might help you to protect your business information.

- If the chosen solution providers and the COTS vendors do not have enough competent resources to set up the COTS systems and integrate them into your IT environment, you will encounter big delays in implementation and increased cost that you cannot recover.
- The installation test performed by you and the COTS vendor is made on the COTS vendor's contractual terms—you get what you see and that is all that the COTS vendor will guarantee, but this does not rule out serious errors seen from your point of view that the vendor looks at in a different way. This can create conflicts, delays, and increased cost.
- If the best COTS solution does not use the same IT infrastructure that you have installed, the acquisition of this COTS system will require that you add to or change your IT infrastructure, which will add costs to not only new technology, but also costs and time to build the knowledge and organization necessary to run the new technology.

Once procured, you are still faced with risk of failure if you do not ensure that your IT infrastructure is properly maintained and supported, especially in the case where your COTS systems change versions, but not at the same time and not synchronized unless you ensure this synchronization yourself:

At one of our clients, we supported the upgrade to a new version of their Client Relationship Management (CRM) system. The CRM system was integrated with the client's compliance control system that was legally required. The new version contained important solutions to legal requirements and had to be in operation fast.

We had established a functional test model that showed that the CRM worked as it should when it was not integrated with other software. However, once receiving data from the compliance control system, the CRM system crashed. The compliance control system had not changed so we claimed that the error came from the CRM system, which the CRM COTS vendor refused to accept.

It took quite a while before we discovered that the crash was related to incompatible data types that had not been a problem until the new version came about. The interface between CRM and compliance control was using a COTS application that was no longer supported in the marketplace.

If the interface application had to be changed, we were faced with several weeks of development.

The original COTS vendor agreement had an escrow clause and the escrow agent could deliver the interface source code to our client.

So far so good, but no one in the client IT organization had a clue as to how to adapt the interface COTS system.

In this situation, we found the names of the programmers who had installed the interface solution. They had established their own company, but they still had expertise concerning the interface COTS system.

It took the experts 20 minutes to correct the situation once on board.

This example shows the importance of the escrow clause, but it also underlines the risk involved with complex IT environments using COTS systems from many different vendors.

Most organizations today are aware of these risks when buying new information systems based on COTS and they respond to the COTS procurement risk situation in different ways:

- Some enterprises develop complete solutions based on a mix of COTS and their own development to be driven in their own proprietary IT environment, where the development and the COTS setup are delivered by turnkey solution providers. This is the typical banking industry solution, where confidentiality issues prevent data from being handled physically outside a country or outside the bank's own IT infrastructure and secure IT environment.
- Some enterprises develop complete solutions based on a mix of COTS and their own development, but this is run by a solution provider in the solution provider IT infrastructure. In this way, the enterprise can concentrate on its information system development and implementation supported by a turnkey solution provider. This can work out well if the IT solution provider is competent and if the information system user organization understands how to manage the IT solution provider, which unfortunately is rarely the case. Some cloud-based or System as a Service (SaaS) COTS-based solutions fit into this category.
- Some COTS vendors develop industry specific ready-to-use solutions that they deliver as turnkey information systems. Smaller information system user organizations can profit from such a solution, but it should not be underestimated that the work with setup and training

of the users still has to be done based on a good requirements specification. No COTS system on its own can be an information system. At a minimum, you need to add company-specific workflows and reporting. Today, this type of solution is typical in the cloud and it is widely used by physicians and dentists, for example, with standard integration and reporting to public health systems.

- Some organizations buy a COTS system, install it, and perform an installation test provided by the COTS vendor in order to release payment to the COTS vendor for delivery of the COTS system once it has been proven to work in the dedicated IT environment. The user organizations then try to manage the setup and implementation of their COTS-based information system by themselves or by using competent resources from a solution provider with expertise and experience in the COTS setup. I have never seen this succeed without major scope changes, delays, and cost overruns in addition to deeply frustrated user organizations, even in the rare cases where good requirements specifications have been developed. The information system solution provider will inevitably accuse the COTS provider of delivering system components that do not work to spec, and the COTS vendor will claim that its system was never intended to be used as the information system solution provider has designed the solution. This is the situation in the bank information system swap case that we will (re)cover next.

- Other organizations procure a turnkey solution to provide them with all the information systems they need. They leave it up to the turnkey solution provider to choose the COTS systems that they need for the solution and, quite often, they also let the turnkey solution provider choose and establish the type of IT environment that the solution requires. This is a very safe way for the solution demanding organization to get an information system solution that is feasible, if the underlying requirements specification is satisfactorily detailed and precise to ensure an acceptable solution, which is rarely the case. This is a very expensive solution, where the detailed requirements are developed by the turnkey vendor and where any change introduced after sign off on the requirements specification costs a fortune to implement. This is very often the situation with information system procurement in the public sector.

4.1 BANK INFORMATION SYSTEM SWAP RISK

Here the purpose is to describe a feasible case of COTS system and solution provider procurement in support of the bank information system swap case.

In this case, we used a solicitation, tendering, and contracting method that could allow us to minimize the risk of delivery failure even though some risks such as the COTS vendor contract terms giving this vendor absolute monopoly already had turned into problems, which would have to be worked around.

Most, if not all, problems with procurement occur because of bad procurement team building and risk management.

While the procurement team has the capability to Plan, Initiate, and Do the buying based on their high organizational level and access to funding, much too often they lack the capability to evaluate the outcome of the buying process.

If the procurement team does not act based on a good requirement specification, there is no way the team can evaluate the quality of what it has acquired. The purchase of COTS system cases that were previously presented are typical examples of this situation:

- The electrical equipment distributor bought a system from a friend. This system is still not an information system and the company is using an increasing number of manual routines and spreadsheets to run the day-to-day business.
- The bank bought a turnkey web-based solution for end user trading on multiple stock exchanges across the world, but forgot to ask the key-stakeholders if this was what they wanted. Unfortunately enough, the users did not turn up to make the solution profitable.
- The bank management bought the same COTS application as a competing bank without considering their specific requirements. The initial implementation project failed completely. This situation could be recovered at a high cost.

The cases mentioned previously show the importance of risk management of procurement.

When you buy products and services from a vendor with de facto monopoly, there is a high probability that you cannot counteract bad performance from this vendor unless you contractually have ensured this opportunity to counteract.

The bank information system swap COTS system and solution provider choice and purchase are a good example of how procurement should not be done.

Bank management bought a COTS application without evaluating the risk from using a de facto monopoly as the provider of the COTS part of their future information system solution. In this situation, the bank management "forgot" to demand an appropriate Service Level Agreement (SLA), which resulted in the COTS vendor delivering solution components with errors that were so serious that it delayed the implementation process for several weeks.

That the bank management also acted without a requirements specification did not improve their risk situation.

In the case of human resource-based service procurement, the quality of what is procured is complicated to evaluate:

- You can get a list of skills, but you cannot see if the person knows how to apply the skills, that is, the competence of the person even in the case of certified resources.
- You can get a list of experiences, but you cannot see exactly how this person performed during this activity.
- Some industries such as atomic power plants or oil production platforms require certified resources for production and maintenance, but even among certified persons you will find very different productivity and delivery ability.
- Productivity of human resources differs enormously because the personality is playing a role, for example, some persons want to deliver perfect results irrespective of the time involved, while other persons deliver a feasible result fast.
- You might obtain a guarantee of results delivered by the procured human resources if you demand it and if you have produced an appropriate requirements spec that focuses on results rather than on resource qualifications.
- It might take a long time before you discover that the procured resources do not deliver what is needed if you do not manage the quality of delivered results very early in the delivery process.

4.1.1 Procurement Risk Mitigation

In the case of the private bank information system swap, a number of external resources had been working on the project for more than 18 months without any progress.

These resources had been delivered from three resource providers:

- One person from the COTS vendor
- More than 10 persons from a worldwide renowned consultant provider
- One person from an agent selling "heads" based on their CV

The bank had not established a procurement team and it had not performed risk analysis of the procurement situation.

This lack of procurement team building and risk management is probably the most important cause of the encountered problems:

- The project manager had simply demanded a gap analysis based on AS-IS and TO-BE observations of the current banking solution components—an impossible task to solve without requirements specifications. The future solution was not defined on a level where TO-BE could be compared with AS-IS. Nevertheless, the resource provider sold resources in bunches to do the job and to be invoiced based on delivered labor-months without responsibility for the result of the gap analysis.
- There was no requirements spec telling the external resources what to do, how to do it, and most importantly why they should do it, so the external resources worked by intuition using whatever skills they had or could get from textbooks on gap analysis.
- There was no competence expectation to the external resources except for their ability to do gap analysis, so they were all more or less junior management consultants without specific COTS implementation experience.
- The resource provider invoiced by delivered labor-months, not by delivered results.
- The project manager never questioned the methods used by the external resources.
- An external resource was recruited from an agent that provided only one guarantee of quality: "The bank could cancel the resource contract with one month's notice." Here the bank was fully responsible for managing the external resource, who was later replaced by a qualified employee.
- The COTS support person offered part time by the COTS vendor had very little COTS application knowledge and only reported documented problems back to the COTS vendor for correction.

This reporting was of such bad quality that problems most often expanded instead of being solved. The COTS vendor had no further contractual obligation to deliver qualified resources for solution implementation.

- No one had thought about producing a requirements specification because everyone thought that the COTS application would provide a solution once the data from the old information systems had been transferred to the COTS application and that the satellite information systems had been integrated with the COTS application.

If the bank had established a qualified and competent Strategy Governance Team to handle the information system swap, it could have eliminated most of these problems based on risk management.

One obvious response to the above risk situation is to transfer the solution delivery responsibility to the vendors of COTS and to potential solution providers because the bank had no experience in COTS implementation at all.

This risk transfer is not something that happens by itself because both types of vendors always act based on standard lawyer-produced contracts that explicitly place all solution delivery responsibility on the buyer.

The standard vendor contract explicitly limits the responsibility of the vendors to deliver their COTS product and listed resources as they are. They leave it up to the buyer to define the acceptance criteria and to control and evaluate that delivery takes place as agreed, a competence that the buyer most often does not possess.

Many organizations do not take up the fight in order to ensure the risk transfer because they lack tools and competence in this sort of negotiation. In this situation, they revert to buying turnkey solutions, where they leave the negotiation responsibility with the turnkey solution vendor.

Turnkey buying does not prevent the need of the buyer to produce a requirements specification if the buyer wants to be sure to get a feasible solution.

In some cases, the production of the requirements specification is done by the turnkey solution vendor, which in all the cases that I have seen has been a pure catastrophe. You need to separate solution quality management from solution production and implementation management. If this separation is not done, you as the buyer have to accept what you get.

The non-separation of quality management from production and implementation was done explicitly by a ministry that wanted to get a fast solution based on pure standard COTS functionality. Their objective was to adapt business to COTS functionality and not the other way around. They thought this would be safer and cheaper.

The ministry wanted to buy an ERP solution as a turnkey delivery. They produced a list of solution facilities that they would need.

Based on this list they asked a few very big COTS-based ERP-solution providers to bid for the delivery of their future COTS-based ERP information system solution.

The public buying process limited the number of possible vendors to five. All of these vendors used the same COTS system for solution implementation. In this case, it excluded a very competent local internationally renowned ERP COTS developer and vendor to be used for the COTS part of the delivery, which resulted in a big loss of local competence development and probably employment.

In order to ensure a satisfactory detailed solution description from the vendors, the ministry allocated €150,000 to each bidder who delivered a satisfactorily comprehensive proposal and solution description (to be their requirements specification) and who did not win the order.

The ministry and the future users were not able to evaluate the details of the proposed solutions, so they contractually committed themselves to accept and adapt their business to solutions that used the standard functionality of the COTS ERP system. This was not very smart because a COTS system does not have standard solutions, only standard functional elements that do not work and do not integrate as an information system before they have been set up to deliver a solution based on a detailed requirements specification. The facility list and the solution provider proposal were far from sufficient as requirements specs.

This solution requirements specification had to be developed with key-stakeholder involvement after contracting. This was a major change to the signed solution provider contract.

The ministry and the future solution users were forced to accept whatever the turnkey vendor delivered. On top of this, the employees had to be trained to work with procedures that were completely new to them.

The implementation became considerably delayed. The cost overruns, especially the additional costs in connection with changes, were enormous.

The complete scope of the solution was continually adapted to what the turnkey vendor could and wanted to deliver, so by definition the delivery of the turnkey solution was a success.

Nobody bothered to ask the users and other key-stakeholders if they were happy because this was, by definition, the case.

The thinking behind this turnkey solution choice resembles the bank system swap case with one important exception. The ministry let the solution provider procure and install the COTS system, while the bank procured the COTS system and thought that they had bought a solution.

4.2 BANK CASE PROCUREMENT RISK RESPONSE

We had to establish solution provider procurement from scratch to ensure the bank information system swap delivery.

The procurement process demanded creative contracting and tendering terms in order to cope with the known problem and risk situation.

When we looked at the procurement tradition in the bank locally and even centrally, we could immediately understand that their legal condition-based standards were of very limited use in our case.

The problem situation of the bank project was good proof of the weakness of the bank's current procurement policy:

- Human resources were simply bought in bunches to be paid by the hour of work in confidence to that the resource provider would deliver resources that understood the information system swap need and who had knowledge and competence to solve this task.
- There was no way the bank management and project management could control progress and delivery quality because there were no requirements specification to compare deliverables against.
- The resource providers were not committed to deliver results, only person-hours. Result responsibility was solely on the shoulders of the Project Manager and the IT Director.
- Money poured out of the bank based on used person-hours that did not deliver any usable solution to the bank.
- The resource provider and the COTS provider used several weeks of bank employees' working time to train them in COTS functionality that they would never use, and the bank had to pay for this service.
- Almost two years were spent by the contracted human resources delivering only more problems and cost.

When this situation was closed out and the project declared in serious trouble, the procurement of additional COTS systems and important IT infrastructure components had not even been initiated.

Solicitation of the corresponding delivery opportunity situation had not been initiated either, which added a new problem because there were long lead times on important IT material that could delay the project.

The relatively complicated quality evaluation concerned with human resource service procurement demands that you perform risk management in order to obtain the following:

- The resources can deliver results of the quality that you need and demand.
- The aleatory resource availability is dealt with and the availability is controlled to be satisfactory.
- The solution provider guarantees not only the quality of the resources, but also the quality of the results delivered.
- You can manage the quality of delivered results from the human resources delivered by the solution provider.
- The external human resources are motivated to work in a proactive way with the internal employee resources.
- Internal employee resources work in a proactive way with the external resources.

These risk management objectives seem obvious, but much too often the risk management processes are not performed at all such as in this case.

For the core-banking information system swap, we would need solution providers and sub-contractors to perform the following jobs:

- Set up the core-banking COTS system in support of the banking business.
- Set up the COTS system-based reporting in support of:
  - Client reporting
  - Risk reporting
  - Public authority compliance reporting
  - Corporate reporting
- Transfer live data from the old core-banking system to the new COTS-based one without losing information.
- Integrate the core-banking COTS system with a future financial control COTS.
- Integrate the core-banking COTS system with a future COTS system for reconciliation handling.

- Integrate core-banking COTS with a fund administration COTS under implementation.
- Build the internal competence required to operate and use the future COTS information systems.

Fortunately enough the bank was not the only bank to use the core-banking COTS. Consequently, the COTS vendors of systems to be integrated with the core-banking COTS had a positive interest to succeed with this integration.

On the other hand, the core-banking COTS system vendor had a very bad reputation as an integration partner, which the bank could do very little about because it had already signed up for both COTS system and system support without an appropriate SLA to oblige the core-banking COTS system vendor to cooperate.

The initial project was replaced with a program managed by an appropriate Strategy Governance Team supported by a Change Management Board. The first task of this Strategy Governance Team was to perform the initial PQA process and to establish the Workgroups to create the missing requirements specification.

While these newly established Workgroups were struggling with the documentation of bank business workflows to be used for requirements specification, the Strategy Governance Team was "converted" into another Workgroup to execute the procurement of competent solution providers and sub-contractors for delivery of solution components, outstanding COTS systems, and missing IT infrastructure components.

In the beginning, we had thought that we just had to find COTS systems for finance and reconciliation, but it soon became evident that the core-banking system was delivered without a satisfactory reporting facility. The core-banking COTS system only handled transactions and a not too efficient integrity of its DB2 relational database. Reporting was the responsibility of the user organization based on a not very user friendly Application Programming Interface (API) to the COTS database.

Consequently, on top of the solution provider procurement we had to procure:

- A financial control COTS system
- The setup of a reconciliation COTS system
- A reporting COTS system related to the core-banking COTS system

The new COTS system setup would have to be done in close relationship with the core-banking COTS system setup, so we decided to integrate the additional COTS system procurement with the solution component procurement in such a way that we could ensure IT, COTS system, and implementation resource availability for a very fast process with many parallel projects (Workpackages).

We were aware of the increased risk from many parallel projects, which we mitigated by building autonomous Workgroups for delivery of feasible results.

The total procurement activity gave us an opportunity to buy additional COTS systems that could be delivered with setup resources that had experience from integration with the core-banking COTS system.

In the luckiest case, the new best buy COTS systems had already been integrated with the core-banking COTS system at other clients; in the worst case, this was not the case and we would have to run the risk of failure if the COTS system implementers could not match the bank solution requirements.

In order to mitigate the worst-case risk, we would try to commit the COTS system vendors to deliver both COTS system and full integration of the COTS system with the core-banking COTS system, that is, turn the COTS system vendors into solution providers.

As mentioned previously, we established the Strategy Governance Team as the Workgroup for procurement of missing competent resources, COTS systems, and outstanding IT equipment. This looks very much like a centralized ivory tower-based buying organization without much contact to real banking needs. We were aware of this organization-based risk situation and did our best to avoid it by doing the following:

- We got the users and departmental managers involved with requirements specification in the form of Workflows.
- We visited the real implementers and their project managers in other banks approved by our and their management to understand the risk we were facing with the chosen and potential COTS applications.
- We asked the reference banks about their experience with integration of other COTS applications.
- We performed an appropriate solicitation of COTS and equipment vendors and solution providers in order to establish a list of competent potential partners although we had very little time to act.
- We never panicked because we knew that one more bad selection of a partner would cause another disaster.

- We prepared a serious and dedicated call for tender that fully reflected our immediate needs and motivation to succeed.
- We prepared all contractual terms beforehand. These terms would be able to ensure a win-win situation and could potentially transfer all possible solution delivery responsibility to the vendors.
- Contractual terms and scope were not negotiable in order to ensure synergy and proactive attitude from all partners and bank employees and managers.
- Change management was ensured in a way that could minimize conflicts except for the conflicts arising out of the missing SLA with the core-banking COTS vendor.
- Only price and time were negotiable once the vendors had presented their accepted preconditions.
- All test cases were developed and agreed to by both internal Workgroups and the corresponding solution provider teams before test runs, so there was "no excuse for failure."
- Payment could only be signed off once the end users had approved the solution delivered based on simulated and final Accept-Testing.

We had to develop a new set of contractual terms that were very different from European public and private standards.

European public and private sector contract standards have been developed by advocates working with the IT industry and IT user organizations with a view to protect primarily the buyer. This is understandable in light of all the failed deliveries and big scandals that you will find in the press.

As mentioned previously, the IT industry seller contracts were completely opposite and removed all result responsibility from the vendors. If the future user organization did not like what they got, as this was documented in the COTS user guide it was just too bad for the user. Error correction time was the sole responsibility of the vendor.

The problem was that such contracts did not in themselves present a guarantee of success. Any vendor could close his shop and restart business under another name. Already paid advances were lost and the public or private clients never got their solution on time.

In order to avoid another disaster, we decided to formulate other types of contractual terms that committed both buyer and vendor to do their best. These terms focused on common responsibility to succeed and placed just as much pressure on the buyer organization as on the vendor organization to allocate the best possible resources as required and agreed. You could say that these contractual terms reflected the "no excuse for failure" principle.

All this preparatory work took almost three months, but the investment paid off in the end. The bank never lost a partner in the process and the teams delivered a feasible solution.

Optimization was never addressed, only feasibility.

4.3 THE REQUIREMENTS SPECIFICATION

Whether you want to develop or to buy your future information system, you are obliged to provide a specification of requirements.

There are two basic forms of requirements specifications concerned with development and implementation of information systems:

1. A list of functional, technical, and general features to be delivered by the contractors without specifying how these features must be established—you express what and the vendor offers how. You might have established a weighting system to the requirements because some requirements are more important than others are. You might also rate the how functionality offered by the vendor.
2. A detailed design of the future information system solution to be set up by the contractors constrained by your technological infrastructure. You rarely find such requirements specifications unless you want the vendor to develop the solution from scratch. On the other hand, your business might be so special that it is the best solution to ensure your benefits.

In the case of the bank information system swap, the second form, relying on documented Workflows, was chosen because:

- A list of features would make the spectrum of solutions too broad with the risk that the bank got a new solution with heavy training needs and the needs of cultural changes that were not demanded.
- Workflows could ensure that bank working procedures would be maintained in the new solution, minimizing learning needs and optimizing Accept-Test performance.

- Workflows resemble more a design than a list of features, but they are still open to COTS-specific user interfaces and reporting.
- The implemented solution had to be established on COTS technology.
- Integration had to be programmed from scratch using among other tools the COTS system API.
- Reporting was very specific and demanded detailed report design.

On this background, it was decided to request competent people for solution delivery before equipment and COTS application delivery.

The chosen vendors and solution providers in all cases could choose their preferred COTS application to work with, except for the already acquired core-banking COTS application.

We wanted to be able to choose more than one vendor and solution provider in order to ensure an ability to get the most competent resources on board.

We knew that having more than one vendor and solution provider left us with a management task of important dimensions, but on the other hand, we had less risk of failure than with a turnkey delivery that we could not control.

In the end, the proposals that we could obtain would limit our choice under all circumstances.

4.4 SOLICITATION FOR COMPETENT CONTRACTORS

There were circumstances originating from our talks with the local reference banks that made us somewhat optimistic concerning our procurement strategy with new contract terms, detailed functional requirements based on Workflows, and our willingness to work with several vendors and solution providers concurrently:

- Not one single reference bank using the core-banking COTS had good experiences from their implementation irrespective of choice of implementation partner. Here we had an opportunity to do better by using another method than theirs that for the most part had been based on turnkey solution delivery by a major solution provider that in all cases had delivered late with heavy cost overruns.
- We knew that all the big consultancy firms had experience from former implementation of the core-banking COTS application, so

some qualified resources might be available for our Workflow-based COTS setup. What we did not know was to what extent they had competence in setting up the parameters of the COTS application.

- We got a list of potential partners from the reference banks to which we talked.
- We established good relationships with project managers in the reference banks.

4.4.1 The Solicitation Process

Based on the list of potential partners, we invited them to tell us what they could do for us in light of the outlined tasks and Workgroups that we had documented in a PowerPoint presentation.

The three major bank-consulting companies in our local environment confirmed that they would bid if invited. They also confirmed that they had access to available and fully qualified resources to do the job within our very optimistic time frame of nine months from contract signature date.

The provider of the reporting COTS application was more of a solution provider than a vendor of COTS applications. This COTS vendor confirmed that it could deliver the reporting required if the bank could provide internal resources and requirements specifications to work with its report developers. This gave us even more hope because its report developers had profound experience from working with the core-banking COTS that we could drag on in our evaluation of work performed by other solution providers. The vendor's requirements to be able to profit from and share knowledge with internal resources were just sweet music in our ears. At that point in time, we did not know how hard it would be to activate the internal IT resources.

All the reference banks had declared that the build-in reconciliation facilities in the core-banking COTS were unusable. The solution providers that we solicited confirmed this situation, but they could of course offer a COTS-based solution if the bank was willing to run the risk.

The bank already had an external provider of its current COTS-based reconciliation application; but this solution provider had never integrated its COTS system with the chosen core-banking COTS application before. As we had expected, it became very difficult for this solution provider to comply with our contract terms. Nonetheless, they were invited to bid.

The provider of a financial COTS application was very consultancy minded and declared that they would be happy to deliver a solution based on their

COTS application. The final negotiations with them as partner turned out to be quite tough because they tried to avoid taking any risk on the sale of the COTS application and because we would not pay for the COTS application before the fully integrated solution had been Accept-Tested.

We also talked to a few major agents and other classical resource providers with experienced resources in our market, but they only sold heads, not solutions, which ruled them out as potential partners.

The result of the solicitation was a list of six potential partners, all of them committed to bid on some part of our call for tender.

4.4.2 Commitment of Internal Workpackage Workgroups

When we initiated the program after closing the original project, we solicited all internal IT employees who had been involved with the failed core-banking implementation.

To our big surprise, we discovered that the internal IT employees had been trained in the core-banking COTS system together with the future users only. This meant that they had no training or any form of experience with running or setting up the new core-banking COTS system.

This lack of education and experience had to be remedied immediately.

The COTS vendor did not have resources available for this training. The COTS vendor system support personnel were prevented from working outside the COTS vendor premises, and actually no structured training in COTS internal technical facilities was available, only documentation.

After very tough negotiations with the COTS vendor, who claimed that it had delivered according to its contract, we obtained that the bank IT personnel could try out the COTS system at the COTS vendor site with access to limited support during the period we used for procurement of solution providers.

This opportunity was rarely used. We had clearly underestimated the lack of motivation among the IT personnel after the failed implementation project.

After this discovery was made, we simply distributed IT coordinators to all Workgroups with the very precise objective to ensure that IT infrastructure needs were resolved as needed and that the COTS system was installed and running in four environments:

- Setup and interface development
- Migration of new versions

- Testing
- Training

This worked out because of pressure from IT management, Workgroup management, and solution provider team management to get needed test and training facilities set up internally. Solution provider team management knew exactly what was required and the problem with COTS vendor support in getting these requirements fulfilled. They had seen the COTS vendor problems before.

4.5 THE CALL FOR TENDER

The Workflow-based requirements specification made the tendering process much easier to manage than a traditional tendering based on a demand for functional, technical, and general facilities:

- There was no need to prioritize facilities because all Workflows had to be developed and implemented.
- Workflows were classified into Workpackages.
- There was one Workpackage for each banking functional area.
- The potential solution providers could bid on one or more Workpackages to be delivered at a fixed price with a fixed deadline.
- The bank Workgroups would create the Workpackage documentation in time for development and implementation.
- The bank Workgroups would create Accept-Test cases based on the documented setup of the core-banking COTS delivered by the solution providers.
- Payment would take place after successful Accept-Testing.
- A number of Workpackages were reserved for internal delivery to ensure the availability of development, test, and implementation infrastructure for the solution providers in the bank.
- No Workpackage activity was allowed outside the bank premises because Workflows and solution setup were confidential bank property. In this way, we wanted to ensure the proactive cooperation between internal Workgroups and solution provider setup teams, which would not be possible unless all resources worked together in the War Room.

- The bank was committed to set up the development, implementation, training, and test environments needed ("no excuse for failure" was ensured).

The program setup with bank Workgroups, solution provider Workpackages, and Workflows were established according to our Coffee Bean method that you will meet in Chapter 5 under strategic initiative implementation. One of the strengths of this method is that it enforces agile cooperation across development and implementation teams, which is needed if one wants to succeed with solution implementation of high quality.

The example here of Coffee Bean method use is quite typical when the primary task is the implementation of a COTS system:

- The Workpackages represent development work that documents the technical setup of the COTS application and closes out with the full support and operation documentation for IT solution governance.
- The Workgroup Workflows represent implementation work that will close out with user guides for the business usage of the COTS application after completed Accept-Testing.
- The cooperation between bank Workgroups and solution provider Workpackage producing teams represents the project management and quality management activities that produce test setup, initial user training, test cases, and Accept-Testing, which ensure delivery of fully accepted results in a feasible IT infrastructure environment.

The Coffee Bean method process requirements were implicitly applied to the contracting terms that all solution providers and internal Workgroups were demanded to sign off. These contracting terms were included with the tendering material.

These preconditions limited the time needed for evaluation of the proposals to:

- A study of the CVs for each resource to be involved in order to evaluate their competences and experience. We also had a very fruitful dialogue with these resources before they were accepted.
- All involved resources had to be known at all times during the Workpackage execution as they personally had to sign a confidentiality agreement to get the key-batch to enter the bank premises.

- The number of consultants to work on each Workpackage in order to evaluate if the proposal was realistic in light of the number of Workflows and reports to be developed.
- A list of reference banks with core-banking COTS implementation.
- The willingness of the solution provider to use the documentation standard established by the bank for solution documentation, test documentation, and training documentation.
- The willingness of the solution provider to adapt to the method used by the bank.

In this way, we could spend more time to establish win-win contract terms that would satisfy the bank's legal department and at the same time motivate the potential solution providers to bid instead of spending time on condition elements that we would not be willing to negotiate anyway.

Our biggest risk was that we would get too few or no proposals at all. We had mitigated this situation in the solicitation process, but we had not shown all tendering elements during solicitation. However, we did get proposals from all parties, but not to our full satisfaction.

4.5.1 The Workpackage Workflow Documentation

In parallel with the Strategy Governance Team development of a new contract for solution providers (sub-contractors), our Workgroups worked hard to produce Workpackage documentation using a document standard with this content requirement:

Name of Workgroup
    Content showing Workflows covered

Workgroup scope (general introduction)
    Participants' experience
    Participants' key figures (employed since, jobs performed)
    IT systems used
    General comments

For each Workflow:
    The required quality of the products (= the purpose, the client/user)
    Essential business processes/procedures
        Origin of data (internal/external organization, function, IT system)
        Treatment of data (manipulation, quality assurance, and control)

> Data delivered to (internal/external organization, function, IT system)
>
> Suggested improvements to product (e.g., to products such as SHARES, INTERNAL DEALS, BONDS), process, and organization

This document standard was adapted from our Business Analysis (Information Requirements Study) standard to this simplified form.

We wanted to have one complete Workpackage from one Workgroup ready for the tendering process so that the potential bidders could see what the bank was committed to deliver as requirements specification.

After reviewing the Workpackage documentation at the bank, the bidders could add their specific conditions and reservations concerning their quality expectations and need for access to further knowledge and decision making during their core-banking COTS setup activity.

No Workpackage documentation was allowed to leave the bank, but I can show an example (extract) from the "Security Transaction" Workgroup Workflows that gives you an idea about the quality without revealing any secrets or confidential information:

SECURITIES TRANSACTIONS Workgroup Workflows

Workgroup members:

NV (Back office) (Workgroup manager)
HS (Trade Desk)
EF (Back office)
KO (Fund Management)
PJ (Front office)
SS (Finance)
LJ (Cash)

Abbreviations used

AM	Account manager
BO	Back office
CBS	Current core-banking system
CCY	Currency
CIS	Customer Information System
COTS	Future core-banking COTS System
EOD	End of Day
FA	Fund Administration
FM	Fund Manager
FO	Front Office
GC	Global Custodian

GCS Global Custodian System
ICS International core-banking application
IPO Initial Public Offering
OMS Order Management System
PM Portfolio Manager
PMS Portfolio Management System (current)
SSI Standard Settlement Instructions
STP Straight-Through Processing
T Trade Date
TD Trade Desk

This Workgroup defines End-To-End Securities transactions accruing from:

- Shares
- Internal Deals
- Bonds
- External Funds
- Investment Funds
- Pools
- Investment Management Funds
- Short Selling

The Workgroup also handles requirements concerning:

- Securities Static Data and Classification
- Reconciliation
- Prices
- Mortgages
- Holiday Calendar
- Portfolio Management Department Documents
- Investment Department Documents
- Insurance Client Documents

GENERAL COMMENTS

General portfolio information is delivered to PM by CBS.

Transactions, positions, prices, securities static data are transferred from GC to CBS.

PM accesses CIS data (Portfolio names) via data warehouse.

CBS delivers brokers to PM which OMS accesses via a data service.

TD/FO establishes the broker agreements. BO sets the brokers up in CBS upon instruction from TD/FO.

Price catalogue for broker fees is present in PMS. TD is responsible for creation and updates of the broker fee tables. This is valid only for shares and bonds. No broker fee structure is available for external funds.

Price catalogue for clients is present in PMS. Standard dealing fees. AM is responsible for updates according to client agreements. This is valid only for shares, bonds, and internal Funds and Pools. No structure for External Funds.

PMS communicates with GCS.

Client Deals are automatically booked in GCS from PMS.

All clients have their depot number registered in GCS. BO is updating the depots in GCS (direct access).

The client's depot numbers are linked CBS.

After EOD in CB, we receive files with holdings, and transactions (plus other non-transaction related files) and CBS will be updated.

CBS checks for missing cash accounts (or if following the booking of the transaction, the account has been closed in CBS).

Bookings are created and will hit customer accounts, nostro accounts, and profit, and loss-accounts.

Broker SSI are present in GCS.

Settlement details like safe custody a/c, cash a/c, swift code, are all updated in GCS

If for different reasons (FX to be added to the deal, new SSI ...) we cannot use the GCS, the deal can be booked manually.

All broker deals have a shadow booking on a banks depot for security transactions.

The two bank sides of the customer and broker deals are matching each other out.

Following our booking of the broker deal in GCS it will catch the transaction and settle the deal.

Transaction will hit our accounts in GCS and we are going to be debited/credited cash on value date.

We have cash accounts for security transactions in every currency with GCS.

We match nostros (client deal against broker deal) the day after value date.

SHARES Workflow (Executed by AM through the Trading Desk)

A client initiates by himself or by his/her AM to buy or sell a share. The order can be received by phone call, meeting, or fax. FO receives the order, executes the trade through the TD and books the deal.

BO processes the deal and the client gets a confirmation by mail.

The required quality of the products (= the purpose, the client/user)

FO executes the order based on client's requirement and makes sure that commission is charged according to client's agreement. TD makes sure to execute the correct order. BO controls that the transaction has actually taken place and checks correctness of charges.

A 4-eye principle is maintained on all transactions before they are executed.

In case of errors, BO informs FO.

Essential Business Processes/Procedures

Front Office Actions:	Org	System Screen	Day
Client Order: Phone, fax, letter from client or Account Manager calls the client	FO		T + 0
Client Identification: Name or account number	FO	PMS	
Agreement on: Transaction type (Buy/ Sell), share identification, number of shares, price, limits	FO		
Check: Price, news, recommendation	FO	REUTERS	
Check: Cash availability and dealing restrictions	FO	PMS	
Write down: Time, client name, share name, number of shares, price and limits in Private Trading Book (Paper book)	FO		
Order: AM calls the trading desk with B/S, Reuters Code, name, number of shares, limits	FO		

Front Office Actions:	Org	System Screen	Day
Trading Desk Actions:			
Registration: Order is entered manually in electronic order book: Order from, B/S, quantity, R-code, type of order	FO	1. 2. 3.	
Check: Market and News—not a technical factor for the order itself	FO	REUTERS and SIX	
Order: Trader places the order by phone with the established brokers (list of brokers in OMS)	FO		
Execution: Broker confirms execution by phone	FO		
Registration: TD registers the execution of the order in the electronic order book	FO	OMS	
Execution: TD informs AM by phone about execution: price, dates, quantity, broker name, any special fees	FO		
Front Office Actions:			
Check: Name, number of shares, broker name and price in Private Trading Book (Paper book)	FO		
Confirm: AM calls, e-mail, or fax his client with information about execution	FO		

Front Office Actions:	Org	System Screen	Day
Enter the deal requires client account number or name; securities type, transaction type, trading CCY, security name or code, quantity, dealt price, client price, broker name or number	FO		
By default: deal date (today), value date (interaction deal date/ trading CCY), broker charges (fees + tax + stamps), client depot number, commission, cash account number			
TD is in charge of updating broker charges in PM			
If special fees the AM has to overwrite the charges shown by default			
All type of fees and commissions can be overwritten, taken in the price			
Security static data is delivered after the EOD via corporate files to CBS and stored in PM database			
Brokers and depot numbers are delivered by CBS			
Holiday Calendar is maintained in CBS and delivered to PM (today only 1 calendar for cash + securities)			
Validation Errors and Warnings are shown during the booking process			

Front Office Actions:	Org	System Screen	Day
Book the deal	FO	PMS	
Check: All trades are checked in Deal Overview (PM) with the Private Trading Book (Paper book) at the end of the day	FO	PM	
Check: The deals are checked on CBS lists	FO		T+1
Back Office Actions:			
Check: Deal capture in PM (if communication problems PM/GCS the deal will have status "failed" and we have to resent the deal via PM) Other status to look for: "unbooked" (for new security not created in GCS/PM)	BO	PMS	T+0
Book: Manual input of the client deal Change of status from "failed" to "unbooked" and then to "booked" We have to enter exactly the same details in GCS	BO	GCS PMS	
Reconciliation: Broker deals against client deals via access list called "control of holdings" All open position on this depot should reflect only customer deals from the day before (exception: broker deals booked by GCS)	BO	PMS ACCESS	T+1

Suggested improvements to product (SHARES, INTERNAL DEALS, BONDS), process, and organization

The Client should have the possibility to place Internet orders, which should be routed via OMS to the broker.

To facilitate STP the Broker should be placed with our custodian or we should have direct link to the Stock Exchange.

All kinds of orders, whether it is executed through TD or direct with the broker, should be registered in OMS with full log.

Automatic matching of client deals and broker deals in OMS with creation of outstanding list on screen/print. (This functionality could also be in place in COTS).

The possibility to have more broker/client relations than the usual one client deal against one broker deal:

- Broker deal with many client deals
- Many broker deals with one client deal
- Many broker deals with many client deals

The very complete Workflow documentation with a direct link to current application screens, Excel spreadsheets, Access elements, and documents in combination with the bank's committed setup of applications and document copies in test mode convinced all potential solution providers about the feasibility of the solution implementation strategy.

The Workgroup Workpackage review and support persons in the Strategy Governance Team were accessible for further information as needed. The level of competence of these persons contributed to the motivation of potential solution providers, who could smell a good reference to come.

The other Strategy Governance Team members produced the contracting material and made themselves available for negotiation with and supplementary information to the solution providers once they started producing proposals.

The actual tendering material did not comprise any detailed Workflow documentation, but it listed and outline scoped the Workpackages on which the solution providers could bid.

After solicitation, we had three potential solution providers that claimed that they could set up and implement the COTS solutions that we needed. All of them were represented in the local marketplace and could demonstrate competence and experience. We would see from their proposals if they also had the capacity to deliver the tendering requirements.

Although the core-banking COTS vendor claimed their system could handle Asset Management (Funds) and Reconciliation using SWIFT and FIX, all references in the marketplace had confirmed that this was far from true.

Our own investigations of the COTS software quality confirmed this, so we knew that we had to find other solutions to these important Workpackages, although management had already bought the Asset Management module.

The bank already used COTS software for Reconciliation via SWIFT, but this software had never been integrated with the new core-banking COTS. On this background, the Reconciliation Workpackage was included for tendering by external solution providers.

We also distributed the tendering material to a highly recommended solution provider that could do all the reporting in the solution. This solution provider also happened to have a Reconciliation COTS package integrated with the core-banking COTS.

Finally, we included the vendor of the already present reconciliation COTS vendor because the bank had a very good experience with this vendor. On the other hand, the experience also showed delayed and quite expensive solution delivery; a situation that the bank saw as an opportunity to change for the better based on the new solution implementation Framework agreement with an appropriate SLA in the COTS vendor contract terms.

The tendering material was later used again to acquire a new COTS package for Financial Control because the delivery of this internal Workpackage failed.

4.6 THE TENDER MATERIAL

The legal department of the bank proved to be a very valuable sparring partner while we produced the detailed tendering material and the future contract for solution providers. The legal department knew the bank standard contract terms so that we could avoid the most obvious blunders in our specific contract formulation; however, they found a real pleasure in the invention of a completely new contract to comply with the core-banking COTS implementation program and the "no excuse for failure" win-win principle.

The tender material contained all the bank-required contractual conditions in such a way that the solution providers could simply choose the Workpackages they could and would like to implement, offer a fixed price and time frame (duration), and sign the contract on their side. The contract material comprised:

- Workpackages for contracting to be referenced by the framework contract
- General contractual terms
- Confidentiality agreement to be signed by all involved resources and resource management in the solution provider organization
- Proposed Workpackage sheet to be filled in with fixed price and time frame and signature from solution provider authorized management
- The bank IT Director would sign off the Framework agreement with the Workpackages assigned to the solution provider

It was stated in the contract conditions that the bank wanted to have more than one qualified solution provider. In this way, the overall management responsibility was with the bank program manager and all other solution providers would have to cooperate proactively in order to ensure that everybody could deliver feasible solution components.

We expected that destructive conflicts would be impossible because no one could win without the contribution from the others, and because everybody was contractually committed to succeed if they wanted to be paid. The only exception was the core-banking COTS vendor, but this key-stakeholder faced a very strong implementation team with high visibility in the market.

In this way, all key-stakeholders were motivated to succeed.

4.6.1 The Workpackages for Contracting

The Workpackages for contracting were presented in a document, which also described the way to deliver the Workpackage in cooperation with other solution providers, the COTS vendor, and the internal employees of the bank.

As we knew that the COTS vendor would guarantee only the functionality that was documented in the COTS vendor's usage and release documentation, we mitigated the risk of conflicts between the COTS vendor and the solution providers by accepting that the solution providers worked only with COTS vendor approved and guaranteed COTS functionality. This was quite natural because the bank management had already signed off the delivery of this functionality from the COTS vendor.

We knew that the bank had bought only a subset of available functionality from the COTS vendor, but nobody knew what additional functionality had to be bought in order to implement the core-banking solution, except from the future solution providers. The solution providers discovered the missing COTS elements and it cost the bank almost the same price as the original purchase price to acquire the missing functionality, but this came as no surprise to the Strategy Governance Team.

The bank IT Department and the former Project Management had never tested the delivered COTS completely for compliance with the very complex documented functionality and database content integrity delivered by the COTS vendor. This was not practically possible, and it was too late to build a good sampling-based test model.

In order to utilize the COTS vendor guarantees the best possible way, we gave the solution providers authority to raise problem and error reports against the bought COTS components in the bank. It was then the responsibility of the bank program change management board to follow through the error correction with the COTS vendor. Whenever possible it was the responsibility of the solution provider to find workaround solutions if there was a risk of serious delays based on waiting for a COTS vendor solution.

Let us look at the Workpackages for contracting document that was delivered as a separate document for tendering and which shows how the risk situation was mitigated.

1. INTRODUCTION TO WORKPACKAGES FOR CONTRACTING

Workpackages comprise all work to be done in order to replace the current core-banking solution with a new COTS-based core-banking solution that comprises:

- Private banking with Safe Custody and Portfolio Management growing into a Straight-Through Processing (STP) solution.
- Asset Management (Funds Administration, Management, Distribution, and Transfer Agency).
- Derivatives Management.
- Client Reporting growing into a Customer Relationship Management solution.

The Workpackage implementation will be the responsibility of selected solution providers, who have the capacity to deliver the required COTS-based solutions on fixed price, fixed delivery date terms, and conditions.

Any solution provider with this capacity can bid on the implementation of one or more Workpackages.

Any bid must be for Workpackage by Workpackage. The bank will refuse any bid, where the implementation of one Workpackage is conditional on contracting any other Workpackage.

It is viewed as an advantage for the COTS implementation program that more than one solution provider participates in the implementation of Workpackages.

1.1. Scope of Workpackage Development

Workpackages for external bidding comprise:

- Establish and Terminate Clients
- Security Transactions
- Cash Transactions
- Derivatives Transactions
- Portfolio Management
- Asset Management
- Reconciliation
- Corporate Actions
- Tax
- Fees (Income and Cost)
- Control and Risk Management
- Client Reporting

Internally handled Workpackages excluded from external bidding comprise:

- Financial Control
- Infrastructure
- Data Conversion
- Web User Interface
- End User Training

The Workpackages handled internally have already competent attention and will ensure the early availability of the system development, test, and production environments required by the solution providers.

The IT services comprising usage access control to all required development facilities, backup, and operation are made available and will comply with the requirements agreed to in the Workpackage contract.

All interfaces to external applications are developed as part of the Infrastructure Workpackage.

Requirements of data not implemented under other Workpackages will have to be demanded as a change to the then agreed Workpackage solution.

A Microsoft Office environment is available for production of documentation, presentations, user guides, and training material.

An organization structure and an organization specific accounting structure with documented accounting rules is available for referencing from other Workpackages. Further accounting requirements in order to reach the required implementation result can be requested by the solution provider as a change to the available Financial Control solution.

1.2. The Workpackage Product

Each Workpackage comprises the following deliverables:

- Setup and documentation of the relevant parameter tables for bank Workflows—The bank Workgroup, who is responsible for Accept-Testing the delivered Workpackage solution, has documented each required Workflow. A Workpackage example can be reviewed at the bank, but it cannot leave bank premises as it is regarded as highly confidential. Detailed Workflow documentation will be available when a contract is signed.
- Setup and documentation of screen images and user data tables to be used by bank users and in IT Infrastructure interface development— A bank documentation standard is established and can be reviewed at the bank. It is the responsibility of the solution provider to adjust and deliver the required COTS interface for bank Workgroup manager approval.
- Implementation of training material, which will correspond to Accept-Testing procedures and will be used for training in the bank after Workpackage implementation—A bank documentation standard is established and can be reviewed at the bank. It is the responsibility of the solution provider to review and approve the documents produced by the involved bank Workgroups.
- Accept-Testing and sign off—The bank Workgroup will deliver test cases for the Accept-Testing. The solution provider will review and approve the test cases and procedure documentation. A bank documentation standard is established and can be reviewed at the bank. The Infrastructure Workgroup sets up the acceptance test infrastructure. Error reporting from simulated Accept-Testing of solution elements will be the responsibility of the solution provider. Error reports must comply with the bank standard, which can be reviewed at the bank. The solution provider is responsible for maintenance of the error log.
- The bank Workgroup manager signs off the delivery of a Workpackage after approved Accept-Testing.

1.3. The Workpackage Development Process

The work of the solution provider is under full solution provider control.

The bank program manager can at any time stop the Workpackage development and cancel the Workpackage contract if the involved solution provider personnel demonstrate a critical lack of competence. This requires that the core banking program manager in cooperation with the solution provider responsible resource manager (named in the Workpackage contract) conclude that the progress of the work can never result in a delivery on time.

The solution provider personnel will have access to bank Workgroup managers in order to discuss and agree on the solution in development.

Bank Workgroup managers will respect that the Workpackage implementation will be done based on COTS facilities and opportunities only. The solution provider sets up the COTS to correspond with the required working conditions expressed by and agreed with the bank Workgroup manager within the physical and logical constraints of the COTS software purchased by the bank.

1.4. The Workpackage Resources

Workpackage resources from the solution provider must have sufficient private banking or asset management knowledge in order for them to interpret the bank Workflows and to transform these into complete COTS-based Workflows.

The COTS setup must wherever possible ensure that data become valid, consistent, integrated, and safe, and that the same data is registered only once.

Workpackage resources from the solution provider must have sufficient COTS setup experience to be able to work efficiently and to respond fast to user requirements, where fast is relative to the agreed delivery date of the Workpackage result.

1.5. The Workpackage Solution Provider Responsibility

The work of the solution provider is under full solution provider control.

Delivery on time and cost is the full solution provider's responsibility.

All work must be done on bank premises. No setup parameters or any form for documentation produced wholly or in part under the Workpackage delivery must leave the premises of the bank.

The solution provider personnel are required to act proactively in order to ensure that necessary decisions and agreements across Workpackage development teams and bank Workgroups are made early, for example, communication of the definition and usage of static data parameters of relevance to data conversion and/or transaction handling.

Solutions delivered to simulated Accept-Testing (typically complete screen solutions and working conditions for a specific Workflow) must in the normal case be approved at the latest the second time that they are demonstrated to the users for simulated Accept-Testing.

The solution provider will respect that the daily work of involved bank Workgroup managers can take priority. Required agreements and decisions must be documented in minutes of meetings.

Meetings must be prepared by delivering the full decision-making foundation to the bank Workgroup manager before the meeting. A "minutes of meeting" standard is available from the bank.

1.6. The Workpackage Contract Terms

The Workpackage implementation is on fixed price, fixed delivery date terms, and conditions.

The required contract terms are attached to this document as Appendix E.

1.7. How to Bid for a Workpackage

Please fill in the attached contract, sign it, and return it to the bank for acceptance.

In order to document your competence, you must declare relevant references with company name, contact person, and telephone number.

You must provide a list of the consultants (with CV) to be used on the Workpackage work.

Your responsible (resource) manager and the Bank IT Director sign the Workpackage contract.

2. SET UP SECURITIES TRANSACTIONS WORKPACKAGE

2.1. Solution Requirements

The solutions must cover all securities transactions accruing from:

- Shares
- Internal Deals
- Bonds
- External Funds
- Investment Funds
- Pools
- Investment Management Funds
- Short Selling

It must ensure correct usage of:

- Securities Static Data and Classification
- Reconciliation, Prices, Mortgages
- Holiday Calendar
- Portfolio Management Department Documentation
- Investment Department Documentation
- Insurance Clients Documentation

2.2. COTS Component Dependencies and Preconditions

The bank has, to its best of knowledge, purchased all necessary COTS components. If a COTS component should be missing, it is the solution provider's responsibility to inform the program manager and the program change board as soon as possible of this issue in order to avoid any delays of implementation.

Only COTS standard components can be used for the implementation.

It is the solution provider's responsibility to document all bank specific definition of parameters with an explanation about what functionality has been established and why.

2.3. Test Requirements

The Bank Accept-Tests the implemented functionality based on end-to-end tests of the implemented solution, which is documented by the solution provider.

The solution will be tested for own feasibility and for feasibility in relation to transactions, which use its data.

2.4. Acceptance Test Scenario

The bank sets up the required and agreed test scenario. The bank cannot refuse to sign off a solution, which corresponds to the implemented agreed Workflows.

The bank IT manager must sign off the delivered solution to release payment. Each of the following Workpackages has exactly the same implementation conditions.

4.6.2 The Framework Agreement

The Framework Agreement covers the usual legal contractual terms that do not change from one party to another.

There are no special conditions included in this contract part and nobody had any problem signing this.

This agreement is included as Appendix C.

4.6.3 The Framework Agreement Consultative Service Agreement

This attachment to the Framework Agreement was a Consultative Service Agreement that was developed especially in relation to the problem and risk situation of the bank.

There is one Consultative Service Agreement for each Workpackage that has been agreed to be delivered by the solution provider with corresponding deadlines and payment conditions.

This agreement is included as Appendix D.

4.6.4 The Framework Agreement Statement of Confidentiality

This attachment is a standard confidentiality agreement that protects the bank's proprietary right to all developed and implemented solution elements, and their underlying requirements specifications and other bank proprietary documents to be used by involved external solution provider resources.

The COTS vendors add their own confidentiality agreement concerned with their proprietary software. The bank had only a usage right to this software.

4.6.5 Escrow Clause

If the COTS vendor goes broke, is taken over by another legal entity that you have no trust in, or delivers such bad quality that it threatens your solution delivery, you want as an emergency action to be able to get access to and use the source code of the COTS software.

The contractual terms that appoint a third party (an escrow agent) to keep the COTS software source code available for your organization at no further cost under such conditions is called an escrow clause.

The escrow clause will probably not be included with the standard COTS vendor agreement unless you ask for it. The COTS vendor will pay the escrow agent and might ask you to pay for this service.

Although you will probably never use the COTS software source code, it is nice to know that you have access to do this—just in case. Therefore, I recommend that you always get this clause included in the COTS software vendor agreement, which is not part of your Framework agreement, but which normally exists for all COTS software that you have the right to use.

4.7 CHOICE OF SOLUTION PROVIDERS

We distributed the tendering material consisting of:

- Workpackages for tendering document
- Framework Agreement with attachments:
  - Consultative Service Agreement
  - Statement of Confidentiality

We obtained:

- Three proposals for core-banking COTS setup and implementation
- One proposal for reporting
- One proposal for reconciliation

None of the solution providers wanted to bid for the Asset Management Workpackage, which was what we had expected.

The Asset Management Workpackage was never attributed to external parties. It took several years to replace the internally developed functionality that in fact only very limited uses the core-banking COTS data.

4.7.1 Core-Banking COTS Setup Proposals

All three core-banking COTS setup solution providers offered the setup of all modules except for the Asset Management module.

All of them had the capacity to deliver within 9 months, although only one solution provider could prove that they had done that before.

All solution providers would involve at least 10 named resources, and strangely enough, the most expensive solution providers had a smaller number of CVs attached to their proposals than the less expensive one.

When we investigated their proposal for the Client Reporting Workpackage and the Control and Risk Management Workpackage, we could understand by the proposed prices and conditions that the solution providers to set up the core-banking COTS had no reporting tool well integrated with the COTS software. They all agreed to this conclusion.

We could not run the risk that the bank's already weak requirements specification of these Workpackages based on low user motivation would cause heavy delays and present a foundation for future expensive changes.

It was therefore with some excitement that we opened the proposal from the reporting solution provider—and what a relief! This COTS-based reporting proposal showed by all means the solution provider's expertise in reporting from the core-banking COTS system. As this COTS system did not offer a reporting facility other than their standard API, the reporting solution provider had based their report generator on this API in such a way that future changes to this API could be contained without demanding changes to the bank reports in most cases. Furthermore, this solution provider could show examples of all the required report types listed in the bank Workpackages.

We could again breathe and return to the proposed Workpackages from the three other core-banking COTS-based solution providers.

All Workpackages, as well as the Reconciliation Workpackage, had been proposed.

Two solution providers were more than two times the price of the third one. This would normally mean that the third one had misunderstood something, but not in this case.

The two expensive solution providers simply listed a set of CVs and had included the Workpackage cost and time proposals per Workpackage as we had requested, but they did not agree to work the way that was outlined in the Workpackages for tendering document.

They claimed to have a better methodology and documentation standard and that there was no way that they would agree to work under the bank program manager.

Both of them offered program management as part of their proposal, which we had not asked for.

On top of this, they claimed a right to adapt and approve the Workflow documentation that had already been approved by the bank. This point was not acceptable because it was clearly stated that the bank placed the responsibility to do workarounds or ask for COTS changes on the solution provider, not on the bank Workgroups.

If the COTS application could not perform a Workflow, the situation had to be evaluated in a much broader spectrum than simply changing the Workflow.

The third and less expensive solution provider was willing to let the bank provide program management with a change management board and accepted all method terms from the Workpackages for tendering document.

Based on their experience with the COTS Reconciliation facility they could offer a solution here, but they recommended using the COTS API and an external solution provider with easily adaptable Reconciliation functionality based on a separate reconciliation COTS system.

In this way, we got ideal partners to provide a complete solution that could fully replace the old core-banking facility.

The currently used COTS for Reconciliation was already loosely connected to the old core-banking solution and the vendor of this package was willing to deliver the Reconciliation Workpackage. We were happy about this opportunity, but unfortunately enough:

- The offer was too cheap to be realistic.
- The listed resources were already known by the bank and were not acceptable.
- The proposal was based on a new technologically attractive release that was not ready yet.
- The integration with the core-banking COTS demanded this new release.

In the situation we were in, we had no real choice. We signed up for the reconciliation COTS with a vendor who was also the only possible solution provider given that the COTS software was still under development.

An important reason for accepting this risk was that the reconciliation COTS system vendor was committed to deliver the new release to another bank six months before our delivery deadline.

Although we obtained much better delivery conditions than we had before, the choice of this COTS vendor as solution provider to Reconciliation gave the bank a lot of work to do internally in support of this vendor.

With some delay, the bank got a state-of-the-art reconciliation solution that both parties could be proud of and that quite a few other banks now profit from.

All other Workpackages were signed off to the less expensive, but very flexible, solution provider who proved to be an excellent partner.

The major problems were with the internal Workpackage delivery because the IT Department did not think that it had to sign off the internal Workpackages, but that is another story.

4.8 LESSONS LEARNED

I have seen quite a few enterprises dealing with resource procurement just as if they were doing recruitment of employees. This is acceptable in situations where it is difficult to find qualified people as employees or where you need to replace an employee for a shorter period. However, it is not a good idea if you need resources to produce a specific solution to a complicated task, such as the setup and implementation of a COTS-based information system.

A COTS system is not your information system solution.

In order to succeed with procurement, you need to manage the risks involved with this process.

Most, if not all, problems with procurement occur because of bad procurement team building and risk management.

If the procurement team does not act based on a good requirement specification, there is no way it can evaluate the quality of what it has acquired.

If a project comprises many sub-projects and broad organizational involvement, you must establish such a project as a program with a Strategy Governance Team and a competent Change Management Board. Here the solution providers participated on the Change Control Board, which avoided all conflict.

If you are faced with a monopolistic vendor and you have a weak SLA seen from your point of view, ensure access to competent knowledge on your side before you negotiate any further conditions with this vendor. If not, you risk losing your case even in court.

Trust your partners only if this trust is based on a good contractual foundation (SLA). This is the best way to avoid litigation and complete program failure.

Internally delivered Workpackages are agreed to in writing with the internal "solution provider" just like the external ones.

Internal solution provider personnel prove their competence based on pertinent CV data just like the external ones.

Demand weekly tracking of progress from both internal and external Workgroups based on reliable Work Breakdown Structures (WBS) and estimations of time to complete (cost tracking is not an issue if delivery is on a fixed price).

Use a project management COTS system in support of WBS registration, planning, and tracking.

Any program and project is risk to be governed by risk management (PQA) in order to ensure efficient risk response.

Risk originates from solution, process, and organization elements; and risk response encompasses management of the quality of all of these elements.

Do not be afraid of doing what no one has done before if this is the only way you can respond to your risk situation.

Things are never so urgent that you do not have time to do what has to be done, right.

Make sure that all victories are victories for all involved parties.

Get the key-stakeholders activated even though they claim that they have no time to offer. Risk management and Finance management suddenly had all the time required when they were delayed, but it was too late to avoid total solution delay.

Close out with Champagne, also to celebrate important milestones.

5

Strategy Implementation

Until now, we have planned our Strategic Initiatives and we have ensured that there are available resources with the required competences to perform the initiatives and deliver the expected results.

Strategy implementation is doing what has been planned in order to ensure that the Strategic Initiatives deliver the expected benefits to the organization.

The Strategic Initiative Teams know their roles and responsibilities to deliver the expected quality of all agreed results. They have the capacity to fulfill the roles and live up to their responsibility and objectives taking into account the conditions of the Strategic Initiatives.

If the conditions of the Strategic Initiatives change to such a degree that the foundation of the strategy needs to change, that is, a new base line is required, the teams have the tools to act and react fast.

The Strategy Governance Team with the Change Control Board and the Process Governance Teams all know how to initiate and perform PQA to improve Success Factors and change direction of the Strategic Initiatives if and when this is required.

However, how does the Strategic Initiative organization become aware of the need for change?

We need a method to ensure that each activity executed keeps its alignment with the corporate strategy under changing conditions such as:

- Leadership and management change their mind as the strategically aligned activities progress; this means that the activity implementation must be adapted to the new attitudes and maybe changed objectives.
- Initially expected resource availability does not materialize, which creates delays and quite often demands completely different processes to reach agreed solution quality.

- Perceived solution quality changes as key-stakeholders and clients change their mind and get wiser.
- Economic conditions change.
- Legal conditions change.
- The quality of delivered solution elements cannot be achieved.
- The involved resources cannot deliver within the set time frame.

The needed method can survey the conditions of all activities that are relevant to the executing Strategic Initiatives and the Strategic Initiatives to be initiated in order to discover:

- Sponsor and key-stakeholder attitude changes
- Risk conditions or events that have materialized and become problems
- Problems that were not foreseen
- Organizational changes with high impact on team performance
- Processes that do not deliver the expected quality of results
- Resources or organizations creating destructive conflicts
- Produced solution elements that are delivered for simulated Accept-Test with too many errors

The method that we have been using comprises the following elements:

- All Strategic Initiatives produce their results based on a requirements specification, which is sufficiently detailed to guide the Workgroups in their development, implementation, project management, and quality management work. The initial requirements specification will be referenced from the solution design, system, and user guide documentation.
- All Strategic Initiative work closes out with Accept-Testing that proves that the result complies with the latest version of the requirements specification, which is the one comprised by the latest baseline. The simulated Accept-Testing on milestone deliveries (sprints or use cases) ensures that the final Accept-Test reveals no new errors. You can compare this to the intermediate tests and the test flight that airplanes go through. The probability of serious errors found during final Accept-Test is close to zero. The documented result of the final Accept-Test is signed off by the sponsor once it has been approved by the involved Strategic Initiative managers.

- Communication:
  - Progress measured by key performance indicators (KPI) that are explained in Chapter 6 on Strategy Governance.
  - Deviations from baseline measured by KPI, but also measured by result quality, changed conditions, and pertinent events
  - Strategy and Strategic Initiative evaluation with Process Governance Team and with Strategy Governance Team ensuring ongoing sponsor support
- Development of functional (error free) fully normalized databases and processes using agreed technology and ensuring the availability of a feasible information system operation and support. Solution design takes place here, but it is only documented for technical use in support of system operation and support, that is, it comprises technical details that the business users do not need to know about.
- Implementation of fully functional solution elements with availability of business operation support facilities and documentation such as user guides and training material.
- The solution design documentation is used for creation of Accept-Test documentation and business user guide production.
- Project and quality management ensure availability and efficiency of resources for development, implementation, and ongoing SAT of solution components delivered from development and implementation. Project and quality management survey and evaluate the Strategic Initiative progress primarily based on SAT results, but also looking after organization and process issues.

You can view the strategy implementation method from a continuous quality improvement point of view:

- Continuous quality improvement means that we **decide** to do something in the form of Strategic Initiatives to deliver required benefits, we **initiate** the initiatives by planning projects and programs, we **do** what we planned to do while adapting to conditions and events, we **evaluate** progress and results, and we restart the cycle with new **decisions** and PQA and so on.
- We have defined the basic quality object classes of Strategic Initiatives by which we can define and measure the stakeholder needs and requirements, when they are broken down into pertinent

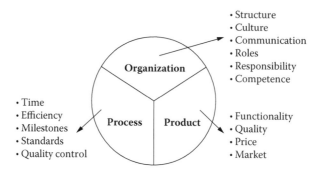

FIGURE 5.1
Strategy quality object classes.

and measurable elements defined by a set of success factors classified into critical success factors (Figure 5.1).

- The method elements that we use to perform continuous strategy quality improvement are the following(Figure 5.2):
  - Process Quality Assurance (PQA) is used for team building, decision making, and planning.
  - The Information Requirements Study (IRS) ensures that the scenario and the solution architecture are agreed to by key-stakeholders on all levels.

FIGURE 5.2
Continuous improvement of the strategy quality.

- The Object Lifecycle Analysis (OLA) defines and documents the detailed solution functionality designed by and agreed to by implementation and development stakeholders.
- Object-oriented functional design ensures an open system architecture that can be integrated with other open systems. The detailed normalization of data and processes ensure solutions that are robust to changing business conditions and requirements.
- Technical design embeds the systems in state-of-the-art technology while ensuring that common standards are applied across all solution elements to ensure the highest possible level of efficiency in construction and test.
- Tests on multiple levels of development based on agile principles ensure fast development, implementation, and adaptation to changed conditions and events. Final Accept-Testing only verifies that already well-functioning solution components work as expected in the future operating environment.
- Systems Management is ensuring the availability and correct functionality of the system seen from technical and business points of view (ITIL and COBIT compliant).

5.1 THE COFFEE BEAN STRATEGY IMPLEMENTATION METHODS

For strategy implementation, we will look at the method elements from a process point of view with development, implementation, and project and quality management that bring us all the way from the original idea and need for change to the final use and operation of the developed and implemented solution (Figure 5.3).

The idea behind the coffee bean strategy of simultaneous Development and Implementation coordinated by Quality and Project Management is to promote systems and lateral thinking and synergy leading to results that exceed the expectations of the stakeholders. It is an agile strategy.

The coffee bean strategy implementation process is divided into the three basic processes of any solution delivery:

- Quality (Project) Management
- Implementation
- Development

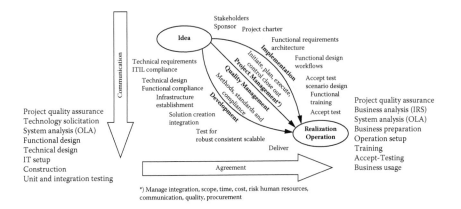

FIGURE 5.3
Coffee bean strategic initiative solution implementation.

The focus is on establishing agreement about what we actually do and deliver under the implementation of the strategy and the Strategic Initiatives once the program and project charters have been signed off.

Development, Implementation, and Quality and Project Management are made up of all required activities, processes, teams, and deliverables from initial formulation of the idea and business case to the final realization of the agreed solutions and their benefits, for example, strategy establishment and governance.

The coffee bean methods can ensure efficient strategy implementation and governance by:

- Offering agility-based processes that deliver measurable results and pertinent information for stakeholder communication supported by the Project Office and the Program Office.
- Usage of processes that are fully documented and easily understood by the involved participants and other stakeholders so that they can review and evaluate the results.
- Fast identification of deviations in cost, quality, scope, stakeholder attitude, and time of all expected results, which allows fast adaptation of the strategy to changed conditions.

The foundation for success with the coffee bean processes is that you visibly involve all relevant knowledge areas in the team building and in the scoping of the strategy with the strategy stakeholders such as we have seen in Chapter 2.

Regular periodic, situation-, and incident-based quality management (QM) activities such as SAT ensure that all decisions about activities, organizations, and deliverables are mutually accepted and signed off by the relevant stakeholders from idea to solutions in business use.

QM activities such as SAT and PQA ensure lateral thinking and synergy. Key development stakeholders meet key implementation stakeholders under (agile) PM/QM coordination in order to develop, test, and verify solution compliance with needs, to establish agreement, and to control progressive elaboration of the solutions to be delivered. This forum is detecting needs for change before it is too late to adapt the Strategic Initiative plans.

The simple agile principle of regular pertinent QM activities is quite often not adhered to, which is the reason for many conflicts and major program and project failures. Regular does not mean periodic unless you talk about a group of managers that meet every Tuesday for a round of golf. Regular in the context of strategy implementation means test and verification with short time intervals "when deliverable components are ready for verification by key implementation stakeholders."

When you hear or read that the communication problems are the primary reasons for project and program failure, it is in most cases based on a lack of regular QM activities. Without these QM activities, you provoke serious risk such as:

- Sponsors do not understand the progress and lose confidence in the project.
- Baseline deviations occur without known reasons, so safe adaption to plans is not possible.
- Developers and Implementers get in destructive conflict.
- Important resources leave the project because they never get the solutions right.

We only measure progress by delivered solution components that have passed final Accept-Test.

We do not want to waste our time in QM meetings just to hear that "We progress well" or outright lies such as "We are 90% complete." We have confidence in our teams and plans until the tracking shows that we are wrong and that we will have to act on deviations from the baseline. When these deviations from baseline occur, we know why and we are able to adapt in meaningful ways.

The QM activities ensure that solution, process, and organization fit together and that they are aligned with stakeholder expectations in both the long run and short term:

- The long run is defined by the duration of the strategy lifecycle (Benefits and Return on Investment and Equity are in focus).
- The short term is defined by the duration of Strategic Initiatives (project scope, deliverables, resource availability, time, and costs are in focus).

The coffee bean processes have been used with success to build a factory, to establish a cash card system, to control offshore oil production, and to swap all applications in a major private bank to name a few rather different program and project types.

5.1.1 Implementation

Implementation comprises all the activities required to ensure that the stakeholders are happy with the solution in full use. The solution can be anything from a new organization, an atomic power plant, the power supply to a factory, a new car, a new bridge, a house, a TV program, or an information system.

Implementation activities comprise:

- Documentation of functional requirements
- Documentation of the outline solution architecture
- Verification of functional and legal compliance
- Preparation and use of Accept-Test scenarios
- Human resource training for use and solution support
- Accept-Test (Simulated and Final)
- Installation and test for operation sign off
- Short- and long-term solution support implementation

5.1.2 Development

Development comprises all the activities required to make the delivered solution available in the quantities and capacities needed to meet the requirements of the stakeholders within the constraints of available and feasible "state-of-the-art" technology and other required resources.

Development activities comprise:

- Development and setup of information system software components
- Unit and integration test
- Validation of documentation of compliance
- Construction of a factory
- Construction of a solution component prototype
- Establishment of the development infrastructure
- Establishment of the operation infrastructure

5.1.3 Quality and Project Management

Quality and Project Management (QM/PM) comprise all the activities that ensure that development and implementation are coordinated and governed in such a way that the activities, processes, and their deliverables meet the expectations of the stakeholders.

QM/PM ensures that both long run and short-term expectations are met by establishing standards and measurable compliance criteria that can make it visible to all stakeholders that the strategy implementation progresses as expected and agreed to.

Communication, team building, decision making, and progress evaluation are essential QM/PM activities.

Other QM/PM activities comprise:

- Coordination of the development and implementation processes
- Simulated and final Accept-Testing
- Evaluation of and response to quality control results
- Documentation of agreement
- Governance support of the project scope and the business model
- Identification and involvement of stakeholders
- Establishment of the project administration with communication management
- Quality management activities
- Quality assurance activities
- Risk Management
- Change management for progressive elaboration

QM/PM ensures that all reviews and tests that are particular to the development process or to the implementation process are handled autonomously within these processes. This concerns development decisions about IT infrastructure and development tool standards, but also the unit and integration tests of a purely technical nature used for bug-free development of solution components. On the implementation side, it concerns the reviews of solution requirement specifications, user documentation, and training material. However, if in any of these cases help is needed from other stakeholders, QM/PM is there to arrange for that in the context of the "no excuse for failure" principle.

QM/PM approves the quality plan and the quality assurance activities in both the development and the implementation process; for example, QM/PM ensures that needed process standards, documentation standards, and open solution standards are available and used. In this way, QM/PM ensures that the involved stakeholders can interpret the results of quality assurance and quality control from any coffee bean process.

5.1.4 Simulated Accept-Testing

Development and Implementation activity is coordinated and aligned in Simulated Accept-Testing (SAT) workshops and through other efficient communication events.

Testing often on real-life solution components is an important part of agile QM/PM.

SAT brings people together from all relevant environments involved with current development, future usage, and cross-functional coordination of solution implementation.

Each SAT workshop covers a defined scenario and scope, where each SAT team member has a well-defined role to play. The objectives of a SAT workshop are defined to be SMART (Specific, Measurable, Agreed, Relevant, and Timed).

SAT makes it possible for the developers to present their solution to the future users and to explain why it is built the way it is. In this way, the SAT process opens up a positive discussion with the implementers and the future users about the quality of solution components.

The SAT participants look for and document required improvements that must be implemented, and they look for potential improvements to be evaluated for implementation by the Process Governance Team.

The open discussion among the SAT participants keeps all doors open; no one says never, only maybe or absolutely yes.

5.2 THE INFORMATION REQUIREMENTS STUDY

The IRS combined with the OLA is used for development of requirements specifications.

IRS documents where to go and how to get there by improving the organization, the information systems, and the corporate processes in light of the corporate strategic objectives and initiatives and the Success Factors established under PQA.

IRS techniques comprise:

- Establishment of a functional breakdown structure of the organization to be studied.
- Establishment of an IRS team to conduct interviews and consolidate the IRS result for Management approval; or, alternatively as we saw in the previous chapter, to facilitate and coach the business users in the development of Workflow-based requirements specifications.
- Selection of people to be involved with the IRS on three organizational levels:
  - IRS Section Management or Workflow Workgroups
  - Departmental Management
  - General Management or Strategy Governance Team
- Interview and interview result documentation standard always asking and answering the pertinent why questions and defining the information objects that flow to, from, and within the IRS section. The Workflow requirements specification is a special version of this document, where the object flow is embedded in the Workflow documentation.
- OLA for IRS result consolidation uses action and structure objects found during IRS interviews to design the outline architecture of the solution to be developed and implemented. OLA is not used in the case of Workflow-based requirements documentation unless there are requirements for improvements to information systems and business processes that must be added to the existing solution architecture for it to be approved by the stakeholders.
- Recommendations with cost and benefit analysis.

IRS provides a foundation for definition of the necessary and sufficient information systems in support of the required business processes in the organization by revealing the information needs from all levels and functional areas of the organization.

The requirements definitions resulting from IRS address both information system-based and organization-based business processes and information concerned with the handling of:

- General management
- Resource management
- Sales orders
- Product development
- Production planning
- Purchase orders
- Accounting
- Logistics
- Customer support

The efficiency and the competitive power of the company are in focus when studying the performance of the business processes and their requirements for information in support of this performance.

The company's IT supported information systems and its use of technology are outline designed under IRS as a solution architecture that can support the company's business performance.

In the case of Workflow-based requirements specification, no outline solution architecture is designed when the Workflow requirements are used directly for solution setup and development in a COTS system that is supposed to be based on a normalized database and normalized processes.

It is fully possible to combine the outline solution architecture documentation with Workflow-based requirement specifications.

When not using Workflow-based requirements documentation, the IRS results in an IRS Consolidated report stating:

- Requirements for enhancements to current business functions
- Requirements for enhancements to current information systems
- Requirements for new information systems
- Required information system architecture
- Suggested development (procurement) of new information systems
- Suggested improvement to technology and IT organization
- Costs and benefits

IRS addresses current information systems problems by suggested enhancements. If the current information systems function well and if they are based on or can utilize "state-of-the-art" information technology, they will normally form part of the future information systems. If not, they will simply be replaced by new information systems. How surviving information systems are integrated with or form part of the future information systems is not within the scope of the IRS. The IRS will define the requirements for integration of information systems, not how it is done.

IRS documentation belongs to the business users who created and approved it supported by other IRS participants, IRS sponsors, and IRS facilitators.

IRS facilitators deliver a documentation standard that makes it possible to compare the study results directly with system analysis and system design documentation in reviews of these and for full traceability of requirements in the developed and implemented information system solutions.

5.2.1 Workflow-Based or Consolidated Report-Based IRS Documentation

When you plan the IRS process, you need to consider the situation of the organization that you will study:

- If the organization has required or developed a new information system and plans to merge its current business processes to be supported by this system, then an IRS Consolidated report is of no or less use because all information objects and their usage are already known. In this case, IRS will establish the organization to create Workflow-based documentation and to ensure a good quality of this for the setup and Accept-Testing of the new information system to be used, managed, and supported in the future.
- If the organization has the intention to improve business process information and information systems and if the organization has not yet purchased or developed a new information system for this purpose, the IRS Consolidated report production is the right choice for IRS because this IRS will allow the organization to choose a future information system solution that fits its needs for business process improvement.
- If the organization has a situation where both types of requirements to business improvements are relevant, then you can use a combination of IRS Consolidated report and Workflow-based solution requirements specification.

Irrespective of your choice of IRS process, the IRS organization building is the same. The different stakeholders simply have different responsibilities and roles.

Workflow-based documentation and IRS section report documentation has a common structure, but IRS with Workflow documentation does not produce management reports and consolidated reports.

The Workflow-based documentation is used directly by solution development and implementation Workgroups that implement the workflows as documented. This ensures business as usual or business conducted in a predefined way for legal reasons such as MiFID and Best Execution in the financial sector. This IRS method allows fast solution implementation if the available COTS systems or other IT-based systems can contain the required solution.

The IRS interview reports are consolidated in order to ensure a consistent solution development foundation with normalized databases and information system processes in support of the business processes involved in solution implementation.

This production of a solution architecture allows the development and implementation Workgroups to be creative during solution production and to deliver solutions that surpass the expectations of the stakeholders.

The normalization of data and processes from consolidation of the section reports supported by the management reports is the foundation for creating improved organization structures and improved communication to achieve the needed benefits of the Strategic Initiatives.

5.2.2 The IRS Organization

A Strategy Governance Team from the level of corporate leadership initiates IRS.

The Strategy Governance Team defines the objectives of IRS and establishes an IRS organization resulting from a PQA process or another decision-making process.

A feasible IRS organization comprises:

- The Strategy Governance Team that has defined the scope, the purpose, and the expected result of the IRS, and that has the power and the budget to allocate business participants and facilitators to the study. The Strategy Governance Team does not necessarily participate in the IRS study work, but it does sign off on the IRS Consolidated report.

- An IRS Workgroup with at least a project manager and a facilitator who establishes the groups to be interviewed (sections, departments, general management) for IRS report production. The IRS Workgroup schedules and performs the interviews and ensures that the interview report is written, approved, and accepted to be owned by the interviewed persons.
- In the case of Workflow-based requirements specification, the groups to be interviewed are replaced by Workflow Workgroups that produce the Workflow-based requirements specification. In this case, the IRS Workgroup facilitates and coaches the Workflow Workgroups with required documentation standards and reviews and formal training in Workflow documentation production.
- An IRS Reference Group represents the knowledge, competence, and experience across the business organization, which is needed for consolidation of the IRS sectional and managerial reports. The IRS Reference Group supports the IRS Workgroup to get the right people selected for participation in interviews and in reviews of resulting reports.
- In the case of Workflow-based requirements specification, the IRS Reference Group is the group of persons that manage the Workflow documentation production and that signs the Workflow documentation off for delivery to solution development and implementation. In this case, there is no IRS Consolidated report produced.

IRS is performed within the business user organization, but the IRS sections are most often a combination of business functional areas that together are necessary to produce a business result.

The traditional form of a business organization is the hierarchy dividing the managerial control of departments covering functional areas such as sales, production, finance, etc.

The market-oriented functional areas can be structured in divisions covering a specific product or a specific market.

The business organization structure is shown in Figure 5.4.

Some organizations are more development oriented than production and distribution oriented, for example, research institutions or dedicated research and development departments in major organizations. The operative areas in this case are most often organized in a matrix-shaped organization with departments/sections in one dimension and projects or products in the other dimension (Figure 5.5).

IRS participants are selected so they cover the necessary knowledge and experience concerned with the business information required by all the

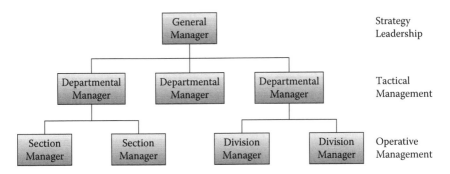

FIGURE 5.4
Classical hierarchical business organization.

involved functional areas irrespective of the organizational structure of the organization to be studied.

The IRS sections often comprise persons from several business sections in order to be able to evaluate cross-organizational information needs and complex business solutions.

A clue to the selection of business function representatives in IRS sections is to look for open-minded positively critical managers with three years' experience in the studied organization or managers with comparative competences, that is, they know and understand the objectives and strategies of their organization, and they have a well-founded idea about how to reach these objectives.

Besides business process competence, IRS demands project management and method coordination delivered by the facilitator and the internal IRS coordinator, who plan and facilitate interviews, make sure that the right questions are asked, treat the answers with critical respect, and create at least the initial IRS results of sectional and management level IRS reports.

	Departments		
Activities	**Administration**	**Marketing**	**Development**
Product A	Resources		
Product B		Resources	Resources
Project 4	Resources	Resources	Resources
Project 7			

FIGURE 5.5
Matrix business organization.

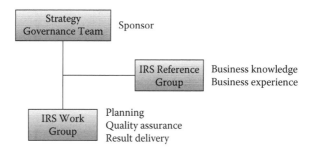

FIGURE 5.6
IRS organization.

A long-term purpose of the IRS organization could be to be a forum for implementation of strategic improvements to business processes, organization, and information systems. It will be a group with a well-established set of methods for cross-organizational cooperation and implementation of business quality improvement (Figure 5.6).

The IRS organization comprises persons such as:

- General and executive managers and their deputies
- Managers from all user departments and their deputies
- Information system managers
- Project Office representatives
- Internal or external facilitators
- Project managers

The selection criteria for the persons are their method knowledge and experience (project manager and facilitator), their business knowledge, their competences, and their experience (information system managers, department/section managers).

The IRS organization can identify who in the organization possesses the knowledge and experience that must be revealed and documented by IRS. The IRS organization does not have to possess the knowledge themselves, it can be a big drawback if they mistakenly think that they do and on this basis write the requirements specification themselves because:

- The real owners of the future solution will refuse to take ownership of the requirements specification.
- Suggestions are passively accepted and pertinent disagreements are ignored instead of leading to better solutions.

The project manager and the facilitator perform the interviews on all organizational levels.

It is an advantage that some interview participants belong to more than one interview group in order to ensure good questions concerned with cross-organizational issues and needs.

In the case of Workflow-based requirements specification, there are no interviews; it is the project manager, the facilitator, and the IRS Reference Group that review and approve the Workflows for delivery to solution development and implementation. Also in this case, it is an advantage that some Workflow Workgroup participants belong to more than one Workgroup.

At least one IRS Reference Group representative should participate in a supportive role in all interviews if the project manager cannot fulfill this role. This representative is supposed to ask the questions not asked by the interviewers in order to get potential hidden facts revealed. He or she will also be able to resolve potential terminology differences between facilitator and IRS interview participants.

In the case of Workflow-based requirements specification, the IRS Reference Group representative will be managing the Workflow Workgroup.

5.2.3 The IRS Process to Produce the IRS Consolidated Report

The IRS process ensures that all possible viewpoints get involved in the study of information systems and information requirements.

The process starts on the sectional level, where business functions are performed for handling of the daily activities of the company.

The approved reports from the sectional level are passed on to the departmental and general management levels in order to prepare the managers on these levels as much as possible before they are studied.

The results from departmental management level are passed on to general management before this level is studied, but these reports are not shown to the sections unless so decided among the interviewed persons.

By following the bottom-up sequence, IRS provides all management levels with the best possible knowledge about the information requirements of the functional areas they control prior to their own participation in the study:

- IRS participants can focus on their proper level of information needs based on inspiration from PQA results and IRS reports from lower or the same levels of the organization.

- Newly appointed managers get to know the organization they manage better.
- Section participants are not constrained by higher management level expectations that they do not know about.

The consolidation of the sectional, departmental, and general management interview reports takes place in a forum comprising the IRS Workgroup and the IRS Reference Group that might have been augmented with new participants during the initial part of the IRS process. The IRS Workgroup and the IRS Reference Group participants divide the work of IRS Consolidated report writing among their participants.

The approval of the IRS Consolidated report and the recommendation it contains is done by the Strategy Governance Team, which might draw the last conclusion on a PQA workshop, where the IRS Consolidated report recommendation is an important input to an improved corporate strategy.

The IRS activities comprise the following:

- IRS is initiated by the Strategy Governance Team appointment of the IRS Project Manager and the IRS Facilitator. The IRS Project Manager and the IRS Facilitator are the IRS Workgroup. The initiation of IRS is often the result of an initial PQA process such as in the case of the military healthcare information system implementation that we will address further.
- The IRS Workgroup supported by the Strategy Governance Team defines the IRS sections. IRS sections are virtual organizational constructs that are required to produce one or more distinct products, which are used by other IRS sections or by parties external to the organization studied. Quite often, the sections selected for IRS interviews and reports will overlap with actual organizational sections such as Order Handling, Major Account Management, Engineering, and Dispatch. Together, the IRS sections cover all business activity in the studied organization.
- The IRS Workgroup establishes the IRS Reference Group of departmental managers to select managers to participate in sectional interviews. Each member of the IRS Reference Group gets the responsibility for one or more sections to be interviewed. It is this member's responsibility to select the persons with the optimal knowledge and experience for interview. A reasonable size of a group for a sectional interview is 1 to 5 persons.

- Interview participant and reference group activation comprise seminars or meetings where the participants from all organizational levels in IRS are informed about what is expected from them and what benefits they and the company can obtain from a successful IRS. Participants in such meetings are prepared by a written introduction to the IRS process accompanied by any available PQA documentation.
- Sectional interviews focus on functions performed in the section and the information used in this context. The interviews result in a section specific requirement specification initially written by the facilitator. The approved requirement specification is formulated as if it has been written by the users themselves.
- Manager interviews and report reviews with the departmental managers and the general managers focus on decision making and the information required in support of this. Based on the dialogues and the business strategy view of the managers, the interviewed managers describe their visions and objectives for the business areas managed by the managers. The resulting IRS management reports are initially written by the facilitator, but it can be written entirely by the managers themselves.
- IRS report consolidation is an interpretation of all sectional and managerial IRS report requirement specifications. Report consolidation results in an IRS Consolidated report that comprises proposals for enhancements to business processes, organization, and information systems. A suggested development and implementation plan for solutions might be included, but such a plan is normally only produced after the IRS closes out in a new PQA process based on the IRS result.
- The Strategy Governance Team (the sponsor) approves the IRS Consolidated report. A PQA process can be used to define the conclusion as a short-term and a long-term implementation plan of the IRS recommendations approved by the sponsor.

5.2.4 IRS Process Duration Estimation

The duration of the IRS will depend on the number of interviews and the possibility of making the right resources available for the IRS. It is likely that the interviews result in requests for more interviews or new scenarios for planned interviews. It is therefore advisable to include some uncertainty in the plan regarding the total IRS duration.

The production of IRS sectional or managerial reports is estimated using a time box approach that limits the time used for interviews and report production:

IRS organization establishment	5 days for facilitator and project manager
Kick-off meetings—one meeting for each 4 interviews	.5 days for all participants in interviews and reviews for production of IRS section and management reports
Interview and interview review including sign off and delivery to the management level above Per section and management report	8 hours for interview 12 hours for facilitator and project manager

The value of the IRS section and management reports is not substantially increased by spending more time on the interviews and the report reviews because detailed description of business processes is not the purpose of the IRS unless Workflow requirements are produced.

IRS studies why the business processes exist:

- Their scope
- Their products
- Their purpose
- Their problems
- Their information requirements
- With whom they communicate

The production of the IRS Consolidated report takes 1 to 3 weeks depending on the number of interview reports to be consolidated, that is, depending on the complexity of the business studied.

The IRS Consolidated report as requirements specification allows information systems to be developed and implemented from scratch because it delivers a solution architecture that allows the designers and developers the freedom to create a solution based on state-of-the-art technological opportunities. In this context, the IRS Consolidated report caters to fast solution development and implementation without sacrificing the solution quality, that is, it caters to more satisfied stakeholders.

The IRS Consolidated report as requirements specification supports the choice and procurement of the COTS systems with the best possible support of the business processes in question.

The IRS Consolidated report solution architecture is an initial scope for COTS setup and COTS-based information system implementation, but it must be replaced by a Workflow-based documentation for final solution setup and implementation. The need for Workflow document production quite often surprises the user organization and leads to unexpected delay. If the Workflow documentation is done by the COTS vendor or an external turnkey solution provider, there is risk of very high cost overrun and delays because the user organization still has to be involved with analysis, reviews, and Accept-Testing.

Workflow Workgroup production of workflow documentation cannot be estimated using a time box principle. The estimation relies on talks with the competent resources to produce the workflow documentation. Only they can tell you how much time they need to produce documentation of the right quality.

The Workflow documentation standard that we presented in the previous chapter facilitates the work, but many other factors such as availability of competent employees, the experience of the project manager and the facilitator to coach the Workflow Workgroups, access to IT-based systems in test mode, and the number of business processes to be documented play a role here.

The Workflow example in the previous chapter that covered only one business functional area was produced in 2 months with very high priority supported by corporate management.

Workflow-based requirements allow relatively fast solution setup and implementation with COTS, but it does not leave much room for business process improvement.

5.2.5 IRS Participant Motivation

IRS participants are motivated to contribute with high quality information in interviews and reviews by giving them detailed information about IRS:

- Why IRS is performed the way it is
- How IRS is performed
- Why the participants have been selected to contribute
- How the participants can contribute
- The benefits to be obtained from the IRS result

The participant motivation assurance activity comprises:

- A workshop where all departmental/sectional managers are invited and introduced to the IRS process and its result. Here they are motivated to participate in the IRS Reference Group and to provide the best possible resources for IRS interviews or Workflow Workgroups.

- IRS kick-off meetings. All selected participants in interviews and Workflow Workgroups are invited to a kick-off meeting. The invitation explains to the invited participants that they have been selected to participate based on their knowledge and experience. Available PQA documentation concerning the IRS is included with the invitation.
- The kick-off meetings can comprise 15 to 20 persons each or a number corresponding to the persons to be involved in 2 to 4 weeks of interviews or to 2 to 4 Workflow Workgroups.

At least the first kick-off meeting should be opened by the sponsor (general manager or Strategy Governance Team member). It makes a strong impression when the kick-off meeting is opened by the sponsor who emphasizes the importance of the IRS and draws attention to the fact that the quality of the result depends alone on an active and committed contribution from the selected participants.

The kick-off meeting comprises a presentation of the IRS process and examples of the documentation that the participants will produce and take ownership of (coached by the project manager and the facilitator).

The IRS process is discussed and it is shown how this leads to a precise definition of the users' requirements for information.

The kick-off meeting duration is 3 to 5 hours depending on the users' motivation and their number of questions during the question and answer session following the presentation of the IRS process.

You should allow 1 week between a kick-off meeting and the first interview. The users are asked to consider the coming interview during this week. They need time to gather documentation and to prepare their requirements.

In the case of Workflow documentation production, the production can start once the kick-off meeting has closed out.

If this motivation assurance is not done, the participants turn up in interviews less prepared, which means that you get less information and in most cases low quality information leading to a long review process with more reviews. You might even get problems to make the interviewed participants take ownership of their report.

5.2.6 IRS Section Interviews

The interviewed persons in each IRS section have been informed in writing and in workshops how to prepare for the interview. They have prepared themselves for the interview by collecting supportive documentation such as reporting examples.

The dialogue in the interview is completely open. There is no questionnaire, but the interviewees know that their IRS section report comprises the following information about the section:

- Scope, purpose, and products of the section
- The purpose of the section's functions
- What would happen if a function was not there
- Information used by the functions
- Information not available which would make it easier to perform the functions
- Information used from other sections and external parties
- Information being filed and maintained by the section
- Information given to other sections and external parties
- How the existing systems should support the functions
- Suggested improvements of information systems
- Suggested integration between IT-based information systems
- Expected benefits from enhanced and integrated IT-based information systems

The section interview uncovers typically 2 to 5 important functions each comprising a number of detailed procedures. The detailed procedures are not analyzed during IRS interviews. Workflows for these procedures will be developed under information system design after the IRS Consolidated report has documented the IRS recommendations. Some procedures might not continue and new procedures might be established based on IRS.

Based on the interview, the facilitator writes the first suggestion for the IRS section report. Before closing out the interview, the facilitator should offer the interviewed persons an opportunity to write the report themselves because they will own the report irrespective of who has written it. Normally the report writing is left with the facilitator. It is important to explain to the interviewee that it is not the facilitator's role to be responsible for the user requirements specification. The facilitator only shows examples and defines a document standard to be used when working out the section report while the users write the content directly or indirectly. The interviewees own their IRS reports.

The user interviews can be extended to or supported by workshops where the interviewees get the opportunity to work with the elements of the IRS documentation in a facilitated environment.

When the interviewees have approved their section report, it is delivered to the IRS Reference Group for final approval. The IRS sectional report is transferred to the departmental managers who are responsible for the section concerned.

5.2.7 Manager Interviews

The section reports are used as a basis for the department managers' description of their information requirements. The department managers often regard the functions of the sections in a different way than the persons interviewed. This is because the functions are performed in view of an operative environment, of which the department manager does not necessarily know all the details.

The department manager's point of view will be influenced by the manager's affiliation with a specific business area such as finance or production. The section reports for this area compared with the manager's ideas and objectives can result in conflicting wishes between sections and department managers.

The conflicts can form the basis for organizational or administrative changes to be carried out in order to utilize the information systems in a better way or simply to enhance the performance of the department and the sections.

To let one of the parties dominate the conclusion is seldom a desirable situation. This is one of the main reasons for starting the IRS on section level without pre-defined limitations in relation to the department manager's point of view.

The facilitator is responsible for the department manager's report complying with the required document standard for this report. The content of the management report is the responsibility of the department level manager and is quite often written by him or her.

Section reports and department reports are complemented with a top management report written by the general manager. The facilitator is again responsible for the report structure, but contrary to the departmental managers, the executive manager quite often wants the facilitator to write the executive level management report. This gives the executive manager an opportunity to understand how the project manager and the facilitator interpret the executive's needs. In this way, he or she can criticize the program manager and the facilitator, which he or she should, and not the other way around.

The executive manager (or the Strategy Governance Team) is interviewed on equal terms with the department managers. It is important that the general manager's report is treated as critically as the other reports with regard to structure and content—it is particularly important that the overall objectives of the company are described briefly and precisely in this report.

The IRS manager interview reports have the following content:

- Scope, purpose and products
- Requests for improved information
- Potential improvements of the basis for decision making
- Expected value of improvements

Departmental manager reports are approved by the interviewed managers and transferred to the executive managers before these executive level managers are interviewed.

5.2.8 IRS Report Consolidation

When all sectional and managerial IRS reports have been approved, we are ready for the consolidation of the sectional and management interview reports.

The complete IRS organization handles the consolidation of the IRS reports.

The IRS organization establishes one or more workshops for final definition of the common terminology and the objects to be used for handling business processes and communication. The IRS consolidation workshops can be done over 1 to 3 weeks, where the duration of each workshop is dependent on the number of interview reports to be consolidated and the number of objects found.

It can be an advantage to have 2 to 3 consolidation workshops with 3 to 5 days between each workshop. This allows the participants to review the conclusions and ideas with colleagues and other knowledge bases between workshops.

In the Defense Healthcare Case, we used five days for one workshop to produce a very comprehensive IRS Consolidated report to be used for procurement of an appropriate information system. In this case, the initial consolidation preparation by the project manager and the facilitator was comprehensive.

Before the IRS consolidation workshops, the facilitator supported by the project manager prepares the consolidation:

- All objects that are the same object, but which belong to different sectional reports and sometimes have different names, are grouped together. A standard object description form is used to define the name, the key, and the reference keys of these objects.
- Each chapter in the sectional report is grouped together with the same chapter from other sectional reports. The resulting cross-section of chapters (with their section ID attached) shows exactly how functions, requirements, problems, and suggested solutions are distributed over the studied organization.
- The cross-section of the input/output chapter is checked for consistency across sections (i.e., what one section delivers to another section should be received by that other section).

This consolidation support material is distributed to the IRS organization before the workshop.

In the consolidation workshop, all inconsistencies and suggestions are discussed. The commonly accepted conclusions are documented in a memo or in minutes from the workshop written by the project manager and the facilitator.

In order to ensure that all IRS Consolidated report writers and reviewers use the same basic terminology and interpret the terms in the same way, the consolidated objects and terms prepared by the project manager and the facilitator are reviewed and approved.

The common terminology and object definitions are documented in standard documents:

- Vocabulary
- Object description

The new consolidated objects are classified into:

- ACTION objects
- STRUCTURE objects

The action objects describe all information needed for performing "time-stamped" actions such as receiving an order, calculating an order, producing to an order, packing an order, or dispatching an order. The ORDER object could cope with this task, but there could also be a PICKING object covering several orders and a SHIPPING object covering several PICKINGS and ORDERS in order to clarify the actions and their interrelationships fully.

For each action object, we want to understand all information needs of each process of the functions handling it and the functions depending on it. Much of this information is kept in structure objects that define INVENTORY, CUSTOMER, PRICE/CONDITIONS, DELIVERY CONDITIONS, and PRODUCT.

The interrelationships between processes and objects are shown in Object Lifecycle Matrices (OLM). The OLA (Object Lifecycle Analysis) establishes the OLM. The OLA technique is also used for data and process normalization in detailed solution design, which is not within the scope of IRS.

For the purpose of IRS consolidation, we perform OLA for high-level functions in order to be able to outline one feasible and complete solution architecture, not for detailed information system design.

An OLM for the ORDER object is shown in Figure 5.7.

The OLM shows for each action object studied what functions are handling it and what other objects are needed by the function for this performance. All of this information is also available in text in section reports and the consolidated report, but the text is less easy to understand than tables and graphics.

The structure of the consolidated report is defined during the Workshop and the responsibility for writing each chapter is assigned to the members of the IRS organization.

Process Object	Order Registration	Order Confirmation	Process …
ORDER	C(reate), U(pdate)	U, D(elete)	
INVENTORY	R(ead)	U	
DELIVERY	C	U	
CONTRACT	R		
CUSTOMER	C,U		
Entity …			

FIGURE 5.7
Object Lifecycle Matrix.

The project manager and the facilitator coach the writing process and ensure that written chapters are distributed to the other members of the IRS organization in time before the closing review.

The IRS Consolidated report is finished by the IRS organization in a consolidation-closing workshop of 1 day's duration.

The IRS Consolidated report documents the information requirements in support of both operative functions and management decision making on all levels of the organization. It shows how the requirements can be fulfilled by a combination of improved business behavior and improved information systems.

The IRS Consolidated report is delivered and introduced to the sponsor at a meeting. After the meeting, the sponsor needs a week to read the report before it is discussed for approval with the IRS organization in a more structured meeting.

The approved IRS Consolidated report belongs to the sponsor.

During IRS, we have produced:

- PQA-documentation and the Activity Description concerned with IRS
- The IRS organization
- Sectional reports with supporting graphics and tables:
  - Object descriptions
  - Data Flow Diagrams (DFDs)
  - Entity Relationship Diagrams (ERDs)
  - Vocabulary entries
- The IRS Consolidated report
- List of Action objects and Structure objects supported by graphics and tables:
  - Cross reference tables
  - Input/output tables
  - Object lifecycle matrices
  - Common object descriptions
  - DFDs
  - ERDs
  - Common vocabulary
- Consolidation memo/minutes
- Reviewed or new PQA-documentation

5.2.9 PQA Closing Out IRS

Once the IRS Consolidated report has been approved by the Strategy Governance Team (sponsor), they can get together for a PQA workshop, where new visions, missions, success factors, and future Strategic Initiatives to follow up on the recommendations of the IRS Consolidated report are established.

5.2.10 The IRS Consolidated Report

The IRS Consolidated report does not form an adequate basis for detailed process descriptions and object definitions, but it makes it possible to describe an outline structure of the needed database content and an outline system architecture, which satisfies the users' requirements for information and functionality in future information systems and business processes.

During the IRS consolidation workshops, several possible conclusions will be evaluated and discussed. The possible conclusions are based on the cross-sectional requirements selected from each sectional report and supported by the management report information.

Only conclusions that are agreed to or which cannot be debated are stated in the IRS Consolidated report. These conclusions can further be listed in a memo or minutes from the IRS consolidation workshops.

Personal conclusions can be listed as ideas for later debate with the Strategy Governance Team (sponsor), which can approve these conclusions for inclusion in the consolidated report or reject them.

A common IRS Consolidated report structure is suggested here:

1	Introduction
1.1	Introduction to why and how the Information Requirement Study was conducted
1.2	The IRS organization with sponsoring Strategy Governance Team
1.3	The Analyzed Organization (visions, objectives, critical success factors, products)
1.4	The section structure used for sectional interviews
1.5	Summary (What can the analyzed organization obtain from improved information systems and business processes?)
2	The company's Information System Situation
2.1	Identified Information System problems
2.2	Requirements to the future Information Systems
3	Business area Information System requirements by business area (not section)
3.1.	General management area
3.2.	Production and logistics area

An outline of a real-life example of a consolidated IRS report is shown in Appendix F.

5.3 THE OBJECT LIFECYCLE ANALYSIS

OLA is used for consolidation of the IRS into a solution architecture that is used during development, implementation, and quality management of the solution components.

The following citation shows why we are using objects and entities in the same way. The entity approach or the extended entity approach is best used for normalization, while the strict object approach is strong when it comes to openness and integration design, and we need both:

> The extended entity-relationship (EER) model is being "threatened" by the object-oriented (OO) approach, which penetrates into the areas of system analysis and data modeling. The issue of which of the two data models is better for data modeling is still an open question. We address the question by conducting experimental comparisons between the models. The results of our experiments reveal that:
>
> a) Schema comprehension: ternary relationships are significantly easier to comprehend in the EER model than in the OO model.
> b) The EER model surpasses the OO model for designing unary and ternary relationships.

    c) Time: it takes less time to design EER schemas.

    d) Preferences: The EER model is preferred by designers.

We conclude that even if the objective is to implement an OO database schema, the following procedure is still recommended:

    1) Create an EER conceptual schema,

    2) Map it to an OO schema, and

    3) Augment the OO schema with behavioral constructs that are unique to the OO approach.

Source: Shoval, P. Experimental Comparisons of Entity-Relationship and Object Oriented Data Models. *Australasian Journal of Information Systems*, Jul. 4, 2007.

We use the EER approach to design and normalize the databases, and we use the relatively simple DFD to show relationships between processes. Our OLM is the textual basis for the DFD.

The solution developing Workgroups continue to use OLA for detailed system design, but here it is combined with agile prototyping, user interface development, distributed solution components development, and simulated Accept-Testing.

The OLA ensures that processes and data are normalized for object-oriented development that ensures integrity and validity of developed solution components in all use scenarios. This method is independent of technology and distribution of functionality. It is used to ensure normalization and integrity of the solution. Once normalization and integrity is ensured, you can distribute the functionality to web, cloud, communication equipment, servers, or portable equipment in search of the best performance.

The distribution of the solution does not change the basic normalized, valid, and consistent solution, it just makes it accessible where data can be handled the best possible way and it makes the solution more user friendly.

Open solutions based on normalized data and processes in standardized IT environments are easy to adapt to new technological and business-based requirements and equipment, and they are easily integrated with other applications such as free and open web services if they are documented with a view to adaptation and integration.

The Defense Healthcare Solution is a good example of this value from OLA. The recommendation is distribution of functionality in order to be able to capture and validate the data with a minimum of human interference, and to be able to use the complete set of data needed in a function that is located geographically distant from the different databases that supply the data.

5.3.1 Matrix Usage

Relationships of different kinds are defined more precisely by a matrix than by textual descriptions or even a diagram. This is especially true if there are many "variables" in the relationship. If you want to show the complete pattern of communication between all sections of the organization, you might want to use an input/output matrix showing for each section along the horizontal axis from which sections along the vertical axis it obtains its information expressed in objects/entities, or reports (Figure 5.8).

The OLA matrix and the PQA matrix are other examples where matrices are necessary to show the full overview of multiple relationships.

5.3.2 Diagram Usage

Various drawing techniques can be used to visualize complicated interrelationships between functions, between entities/objects, and between functions and entities/objects.

I recommend using diagrams and matrices whenever possible because they give a much better overview of complicated situations than even the most precise and well-formulated text. With matrices, it is much easier to check for completeness of relationships such as you have seen with the risk response matrix and the input output matrix.

Entity To Process From Process	Order Registration	Order Confirmation	Process...
Order registration	ORDER		
Order confirmation	ORDER		
Order dispatch	STOCK ORDER	ORDER STOCK	
Contracting	CONTRACT	CONTRACT	
Process...			

FIGURE 5.8
Input output matrix.

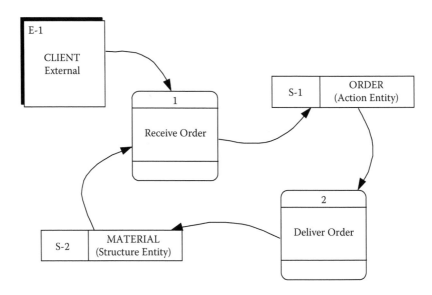

FIGURE 5.9
Data Flow Diagram drawn with SILVERRUN.

Diagrams do not answer the why questions, but in most cases they can illustrate the how. Diagrams do not replace text, they make the text more readable.

The relationships between functions and objects can be shown in DFDs. Please be careful not to complicate the DFD with too many processes and objects. Complex structures are better broken down into small, interrelated diagrams, where each diagram tells a partial story (Figure 5.9).

The relationships between objects can be shown in ERDs. The ERD has the action object in focus and the number of entities should be limited to a selection that is easily understandable (Figure 5.10).

Relationships between structure objects are more often of a parent/child type, where one object inherits information from another one, for example, CUSTOMER inherits the account number from the DEBTOR, and CUSTOMER WAREHOUSE inherits customer name from the CUSTOMER giving the relationships shown in Figure 5.11.

In order to understand all implications of a delivery to a warehouse, one needs information from both customer and debtor. Inheritance and relationships showing, for example, the customer types that constitute the customer object are better described using structure charts such as shown previously.

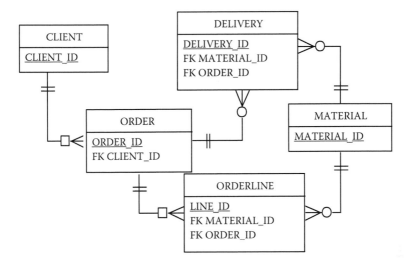

FIGURE 5.10
Entity Relationship Diagram drawn with SILVERRUN.

The diagram in Figure 5.10 tells a story:

* The Underlined item is the primary key
* The items marked with FK are foreign keys, i.e. Reference keys
* Material has zero or many Order lines and Deliveries
* A Client has zero or more Orders
* Orders and Material can be split on many deliveries.

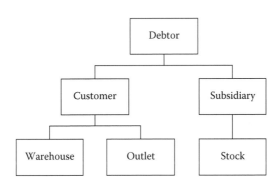

FIGURE 5.11
Hierarchical inheritance data model.

5.4 IRS CASE STUDIES

Two case studies have been included to illustrate the IRS process and the IRS results:

- The Defense Healthcare IRS Consolidated report
- A private bank sectional report

5.4.1 The Defense Healthcare IRS Consolidated Report

The setup of the Defense Healthcare IRS organization and the preparation of the study with PQA have already been shown in Chapters 3 and 4.

The sectional breakdown structure and the development of the IRS Consolidated report are shown in Appendix F, where the IRS Consolidated report structure is shown and explained.

In the consolidated report, the process to produce the report is described.

The sectional and management reports followed exactly the guidelines in Appendix E, but the actual reports are not shown, only the guidelines.

5.4.2 The Private Bank Sectional Report

The private bank sectional report is shown as Appendix G.

5.5 SOLUTION DESIGN AND DEVELOPMENT

Solution design and development are the original subjects for agile behavior. To create agreement between designers, developers, and solution demanding users has always been a challenge for several good reasons:

- The demanding users have not always been able to describe their needs in sufficient detail for the developers and designers to create and deliver what the users want.
- The designers and developers have created solutions that are closer to their technological constraints and opportunities than to the business functionality demanded by the users.

- Solution requirements reviews have been conducted without design and development involvement leading to wrong interpretation of the user demands and a lot of wasted time and money to develop unacceptable solutions.
- The users are too busy to be involved in satisfactory solution Accept-Testing leading to situations where serious errors are discovered under system usage.

We have already seen some methods and tools that can help overcome these problems:

- PQA and team building ensure a safe foundation for establishing agreement about the Strategic Initiative Success Factors.
- IRS with initial PQA establishes a requirements specification for solution development and implementation that fulfills the PQA Success Factors.
- Post-IRS PQA establishes the best possible teams for solution component procurement, development, and implementation.
- OLA with text-based and graphical solution documentation elements offers an opportunity to create easy to use and easily understandable detailed design documentation to be used by both IT experts and solution users.
- The OLA method offers an opportunity for data and process normalization that is required for solution integrity and consistency assurance.
- Workflow-based documentation with access to user interface documentation is a sufficient tool for IT supported solution creation.

While it is relatively easy to get the solution users and their management involved when starting up Strategic Initiatives with PQA, it is increasingly difficult to get the users motivated and committed during solution development and implementation.

Already under the IRS-based solution architecture development it becomes difficult to get the users motivated and involved. Some users fear the future and others are too busy doing their job to play a role in IRS. They participate in interviews because they are asked to do so, but they do not want to write the interview reports or take committed ownership. The users are very often dissatisfied with interview reports written by an IRS facilitator who needs time to understand the users' unspoken requirements—the unknown knowns that all the users know so well that it is not necessary to write them down.

When we arrive at detailed solution design and development, there is no limit to the unwillingness to participate from the business user side. A good example is the private bank system swap that was close to ending up a flop because the business users had no confidence in the development and implementation project—with good reason.

It is because of this natural suspicion and lack of motivation that Strategic Initiatives require that special attention be made to the business users:

- They are coached to produce their requirements with high quality from business procedure improvement to information system solution implementation. High quality means that they are proud of their contribution and take ownership.
- Their involvement is kept to a minimum because they are driving the business that earns the profit to pay for the improvements.
- They are involved in activities that promote their motivation and willingness to participate in the Strategic Initiative.
- They are motivated to participate in activities where their knowledge is absolutely needed by proving to them that no solution can be delivered without their contribution.

In order to protect and support the business users as best as possible, we involve them only when we want them to evaluate finished and technically error-free solution components supposedly responding to their real business requirements and expectations.

The only moment where this is not true is during the IRS interview, where we all start from scratch with the exception of the IRS Reference Group and the users themselves. This process is highly risky and that is why we make a big effort to inform the users about what we expect from them.

This respect for the users and their business work priority is part of the agile principle. When we involve people, we do it in processes where they implicitly will understand that their opinion and judgment counts.

During solution design and development, we leave the object orientation and the normalization with the experts and we depend only on the users to accept their own part of the solution. This happens during simulated Accept-Test and under final Accept-Test.

In the meantime, we use members from the IRS Reference Group and the IRS Workgroup to work directly with the solution designers and developers to develop the solution components ready for installation, operation, and Accept-Test. This is the forum for the original agile principles, but the agile principles also hold true during simulated Accept-Test.

Only completed and technically error free solution components are delivered for Accept-Test.

5.5.1 Design Documentation

The design documentation is structured by the system components structure. It explains how process and data normalization have been implemented. This corresponds to what you can expect from a COTS vendor user guide before the COTS system has been adapted to your business specific information system needs.

This documentation is produced by the solution designers and most often by the database designers among them.

5.5.2 Training Documentation

The IRS Reference Group and the IRS Workgroup produce the training documentation and the training Workflows. Contrary to the user guide, it contains complete business procedure use case Workflows with training data. There is often one set of training material for each type of business solution user, for example, one for the front office and another one for the back office.

5.5.3 User Guide Documentation

The user guide is structured according to use cases. The user guide documents the exact Workflows with user interface interaction and data validation rules that produce the users' business solution results as required.

5.6 SIMULATED ACCEPT-TESTING

SAT is used for meaningful involvement of the future users of an information system solution.

SAT allows the involved parties to verify that their business workflows are fully supported within the scope of the tested solution components.

SAT is performed with intervals of 2 to 4 weeks during the development and implementation of business solutions. The conditions for SAT of a solution component are:

- The solution component works technically and is robust for most thinkable user errors, that is, it works as designed and gives nice error messages when users treat it differently than expected.
- The technical infrastructure works when the users arrive for testing and it continues to work while testing except from unexpected but possible IT breakdowns. It has been ensured that the users can log on to the solution and use it as designed, for example, no one has touched the firewall settings overnight.
- Support from IRS Reference Group and Workgroup and from COTS or other system developers is available in the SAT environment.
- The users have been educated in the usage of the solution component for SAT and they have approved the training material with recommended improvements (on the SAT Issue list). In this way, the involved users know what they can expect.
- The user guide for the solution component is available, but it is not the SAT testers outside the IRS Reference Group and the IRS Workgroup who recommend changes to this document.
- The technical developers have presented the solution and its strength and eventual issues in a presentation in connection with the Accept-Test user training.
- Aside from structured business process oriented SAT test cases that the user is obliged to use first, the users are also requested to test other usage opportunities of their own choice. Sometimes this gives surprising results that can improve both the technical and the business functional solution.
- Error and Issue reporting document standard is available and is filled in by the IRS Reference Group and the IRS Workgroup participants immediately when an error or issue has been discovered. Error descriptions are further copied to the developers for handling.

SAT testing is regression testing because the SAT is repeated until no usage blocking errors are found. Each regression can comprise more functionality, not just corrected functionality.

5.6.1 Test Model Handling

Test models and testing could easily be a subject for another book, so here I will just outline a few principles for a SAT checklist.

Unit testing and integration testing during system development are left with the developers and designers. They simply have to deliver error-free solution components for SAT. All other developed solution elements are of no interest to SAT. This is the agile principle and the foundation for agile tracking of progress.

The developers, designers, and the IRS Reference Group and Workgroup have established the sequence of solution modules (use cases) to be created and delivered for SAT.

I recommend using a system such as HP Quality Manager to keep track of your SAT plan and the SAT progress. When a solution module has passed SAT, it is ready for final Accept-Testing. If a solution component that has been approved from SAT is changed for some reason, it is SAT tested again, eventually together with other SAT ready solution components. This is the "regression testing" principle.

SAT test model checklist:

- Use a test management application to keep track of the test progress. I have used HP Quality Manager with good results.
- The test cases cover all business procedures from end to end.
- Test data, especially structure objects, allow for all result and procedural variants to be tested.
- Document exactly the expected results in the test case Workflow document.
- Enumerate the test cases and register them in the test management system. Personally, I document the test Workflows outside the test management system, but it is up to you how you want to do it.
- Make sure that an error is not based on a test document writing error before you report it.
- Make sure that all test cases are at least outlined in the training material to be used for SAT user introduction to SAT.
- Personally, I register test results outside the test management system, so that this system only tells the users who have tested, what has passed, and what has not passed. Again, it is up to you to decide how you want to handle this.

- Make sure that you have an Issue and Error documentation standard and use it. Issues and Errors are tracked for resolution, for example, such as shown in the next section.

5.6.2 Test Result Documentation

The test is documented with a sign-off sheet that lists what has been tested by whom and what the result was. Attached to this sheet is the test result documentation with error and issue list.

A SAT test can declare a solution module ready for use even though less important issues and errors are outstanding.

For test result documentation, I propose the following standards:

- An error and issue tracking document
- An error and issue description document that allows tracking the error or issue to its source, or even its reason if possible

The document standard that I have used for error and issue tracking is an Excel spreadsheet with the following columns for each error or issue:

- Error or Issue ID that interfaces to the Error and Issue description document
- SAT date
- Urgency (IMMEDIATE CORRECTION, NEXT PATCH, NEXT VERSION)
- Severity (A Blocking, B Major, C Minor)
- Status (NEW, ANALYZED, PENDING, FIX SCHEDULED, SOLVED, VALIDATED, REJECTED)
- Status date

The error and issue description document is shown in Figure 5.12.

5.7 FINAL ACCEPT-TESTING AND SIGN OFF

The final Accept-Test is a mere verification that the solution is ready for use. There are no formal test cases except for the ones used for SAT.

Project ID		SAT Module	
Date	DD/MM/YYYY	Function	Ref. to list
Issued By	Last name First name	Test ID	Name/Version
Solution Needed	DD/MM/YYYY	Tested by	Last name First name

Subject	File, Window, Menu, Report
Appendices Ref	Screen dump, Report, ...

Urgency	Severity Level	Status			
Immediate correction	Blocking (A)		New		Solved
Next General Patch	Major (B)		Analyzed		Validated
New Version	Minor (C)		Pending		Rejected
			Fix Scheduled		

Error Description

Solution Description

Handling

Handling	Person Name, Department	Date
Approved by		DD/MM/YYYY
In solution production at		DD/MM/YYYY
Resolved by		DD/MM/YYYY
Accepted by		DD/MM/YYYY

FIGURE 5.12
Error or issue document.

The solution might go into production with errors and issues still open, but only issues and errors that have no influence on the integrity, validity, and consistency of the solution.

The developers, designers, the IRS Reference Group, and the IRS Workgroup sign off on the accepted solution delivery.

The sign-off document with outstanding errors and issues is stored for reference.

5.8 SOLUTION OPERATION KICK-OFF

Solution operation kick-off is a champagne party with high attention from the future support organization that will be responsible to keep the users happy until new changes are introduced.

If the sponsor does not pay for the champagne, you have a problem.

5.9 LESSONS LEARNED

Strategy implementation is doing the right things at the right time in the right sequence in order to ensure that the strategy gives the expected benefits to the organization.

You can view the strategy implementation method from a continuous quality improvement point of view.

The coffee bean methods can ensure efficient strategy implementation and governance by:

- Offering agility-based processes that deliver measurable results and pertinent information for stakeholder communication supported by the Project Office and the Program Office
- Usage of processes that are fully documented and easily understood by the involved participants and other stakeholders so that they can review and evaluate the results
- Fast identification of deviations in cost, quality, scope, stakeholder attitude, and time of all expected results, which allows fast adaptation of the strategy to changed conditions

The truth is that things never turn out the way we planned them unless we adapt our plans to what we actually do and produce. Communication is crucially necessary here.

We measure progress by delivered solution components that have passed final Accept-Test only. We do not want to waste our time in QM meetings just to hear that "We progress well" or outright lies such as "We're 90% complete." We have confidence in our teams and plans until the tracking shows that we are wrong and that we will have to act on deviations from the baseline.

The QM activities ensure that solution, process, and organization fit together and that they are aligned with stakeholder expectations in both the long run and the short term.

Implementation comprises all the activities required to ensure that the stakeholders are happy with the solution in full use.

Development comprises all the activities required to make the delivered solution available in the quantities and capacities needed to meet the requirements of the stakeholders within the constraints of available and feasible "state-of-the-art" technology and other required resources.

Quality and Project Management (QM/PM) comprise all the activities that ensure that development and implementation are coordinated and governed in such a way that the activities, processes, and their deliverables meet the expectations of the stakeholders.

IRS documents where to go and how to get there by improving the organization, the information systems, and the corporate processes in light of the corporate strategic objectives and initiatives and the Success Factors established under PQA.

The IRS provides a foundation for definition of the necessary and sufficient information systems in support of the required business processes in the organization by revealing the information needs from all levels and functional areas of the organization.

A clue to the selection of business function representatives in IRS is to look for open-minded positively critical managers with 3 years' experience in the organization or managers with comparative competences, that is, they know and understand the objectives and strategies of their organization, and they have a well-founded idea how to reach these objectives.

A long-term purpose of the IRS organization could be to be a forum for implementation of strategic improvements to business processes, organization, and information systems. It will be a group with a well-established set of methods for cross-organizational cooperation and implementation of business quality improvement.

OLA is used for consolidation of the IRS into a solution architecture that is used during development, implementation, and quality management of the solution components.

For the purpose of IRS consolidation, we perform OLA for high-level functions in order to be able to outline one feasible and complete solution architecture, not for detailed information system design.

The IRS Consolidated report documents the information requirements in support of both operative functions and management decision making on all levels of the organization. It shows how the requirements can be fulfilled by a combination of improved business behavior and improved information systems.

The IRS Consolidated report does not form an adequate basis for detailed process descriptions and object definitions, but it makes it possible to describe an outline structure of the needed database content and an outline system architecture, which satisfies the users' requirements for information and functionality in future information systems and business processes.

The design documentation is structured by the system components structure and explains how process and data normalization have been implemented.

The user guide documents the exact Workflows with user interface interaction and data validation rules that produce the users' business solution results as required.

A SAT workshop covers a defined scenario and scope, where each SAT team member has a well-defined role to play. The objectives of a SAT workshop are SMART (Specific, Measurable, Agreed, Relevant, and Timed).

SAT allows the involved parties to verify that their business workflows are fully supported within the scope of the tested solution components

Unit testing and integration testing during system development are left with the developers and designers. They simply have to deliver error-free solution components for SAT. All other developed solution elements are of no interest to SAT. This is the agile principle and the foundation for agile tracking of progress.

6

Strategy Governance

We have planned and initiated our Strategic Initiatives with a solid requirements specification and documented result expectations agreed to by sponsor and other key-stakeholders, but how can we make sure that we actually get what we want?

What we demand and expect is supposed to happen in the future, but we know an old saying (origin unknown) that "It's hard to make predictions, especially about the future."

We might get help from pure luck by hanging a horseshoe over our door such as did Niels Bohr, a famous Danish scientist and Nobel Prize recipient, who replied to a friend that asked him if he really believed that it helped: "Of course not... but I am told it works even if you don't believe in it."

Our plans and our requirements specifications are predictions and it is pure luck if we get a solution that fits the requirements specification—unless we manage the solution delivery.

Contrary to good weather and true love, which we all want to achieve but where we have very little influence, the delivery of business solutions is after all more under our own influence:

> Within the budget and other constraints concerned with, for example, people, technology, and legal compliance, we can achieve a higher probability to get what we want if we make an effort to manage the implementation of our required result.

Strategy governance is performing this management task.

While projects are relatively easy to manage because they produce predictable and tangible results with known resources and well-defined baselines, once they have been planned and are executing the program part of Strategic Initiatives is more difficult to manage because the benefit target is much less tangible.

The Strategic Initiative not only plans and executes projects; it also changes business behavior, re-organizes organizations, and invests in the development or acquisition of new products and improved methods where the outcome depends on how the market and the internal stakeholders accept the change. This makes the outcome of Strategic Initiatives random, and for both forecasting and tracking we need to apply statistical methods in order to understand the magnitude of variance with which we are faced.

In the Strategic Initiative world of random outcomes, agile planning and tracking of our activity allows an easier to understand follow-up on progress. By measuring progress only by user and management accepted solution components delivered, installed, and ready for production, we know what we have and we have a steadily improving foundation for estimation based on the gained experience.

To work in an agile way does not happen without prior planning and organizing. Solution components that lend themselves to agile processes, organizations, and management are identified and you plan for their scenarios to be available with "no excuse for failure."

The benefits from agile solution delivery in terms of solution quality, process efficiency, and cost reduction far outweigh the planning effort and the scenario establishment investment for their realization.

Agile planning and tracking do not completely remove the uncertainty from your planning and execution of Strategic Initiative solution component delivery:

- We cannot be sure that we will not have to add use cases or other unforeseen components to our solution before we can make the stakeholders happy.
- Even the best agile team can lose important members.
- We do not know if the implemented solution component will yield the expected benefits, especially if we have not estimated or forecasted these benefits in tangible measurable terms.
- We are still depending on the forecasted figures of duration time and cost to completion of the outstanding solution components from the Workgroups being reliable.

In order to satisfy the stakeholders, all Strategic Initiative elements of solution, organization, and process stay variable and manageable until the day when the Strategy Governance Team decides that the program can close out. These are the Strategic Initiative elements (variables) that the involved organization of Governance Teams, PQA decision-making teams, and Workgroup Teams can manipulate in order to ensure that all stakeholders accept the final result.

Whenever it becomes evident that the original targets cannot be met or that the original targets no longer are valid, it is time to change the strategy. Changes and adjustments to solution scope, organization, and process make the Strategic Initiatives fit the reality that becomes visible only as the Strategic Initiatives progress and meet the obstacles that we did not expect and for which we did not plan.

6.1 NEGOTIATION

Strategic Initiative governance is about negotiation.

The Governance Teams and the Workgroup Teams continually get and deliver information that indicates that planned activity is not executing the way it was planned:

- Resource availability is not as promised.
- Resource skills and experience is not what was contracted, so important activity does not yield the result quality expected and it takes too long to get the results.
- Critical activities cannot start because they wait for resources to be released from other activities.
- Critical activities are interrupted because important resources leave or because development components from the COTS vendor do not work as expected.

I have never been involved with projects and programs that do not have these problems. The risk management can just tell you that it might happen and that you need to plan for this eventuality with appropriate risk response. However, planning to avoid these events is only possible if you double or triple the resources and with them the cost of the project and the required budget.

By allowing a little more time for delivery, you might have time to adapt to the problem without major cost increases and still have happy stakeholders.

These alternative possibilities for avoiding or mitigating the risk demand that the Strategic Initiative Sponsor and Governance Teams negotiate with solution delivery Workgroups of internal and external resources in order to find the best possible way to handle the risk situation.

It is never a good idea to let one party dictate how to mitigate the risk. This way of tackling the problems removes responsibility from parties that might have good ideas and might be willing to take on responsibility.

Negotiations without one dominating party is a first step on the way to obtain win-win solutions and synergy effect, while domination from one party has a demotivating effect that most often touches all involved stakeholders because the dominating party will have doubts about the effect of what was dictated in the light of the demotivation of the dominated party.

To prepare for negotiation, we need to know where we are and we need to agree on where to go based on what we know now about the Strategic Initiative conditions.

When a major SAP client chose an implementation partner for its future Information System solution delivery, the client thought that SAP had this solution embedded in all of its functionality.

On this basis, the client committed to accept solutions based on what SAP could offer as standard only. This implied that the client accepted to adapt the client organization and the client business process workflows to whatever SAP could offer. In this way, the implementation budget was kept to a minimum.

Unfortunately, SAP did not have all the solution elements foreseen by the client, so a lot of adaptation was needed. Certain legally compliant solution elements could only be delivered at costs that far exceeded the initial budget. Alternative solution elements existed in SAP, but setup requirements already implemented in other solution components prevented the client from using these solution elements.

The solution delivery partner was very clever at negotiation, so it was able to convince the client that its original scope and baseline was not what it really needed.

The client management was under pressure to get the solution implemented as fast as possible, so it was very happy to agree with the solution delivery partner about the "minor" scope changes.

Twelve months later client management and the solution delivery partner could agree on and announce publicly that the implementation program was a success, delivered on time and on budget. What they did not tell the public was that the success had been achieved on one-third of the solution elements expected by the client at the start of the solution implementation process.

The third of the expected solution that was implemented was in production and the users of this part were happy.

This story is an example of how you can make stakeholders happy through Strategic Initiative governance:

- Whatever you think you need or whatever you require from the outset of the Strategic Initiative is never what you get.
- You are successful if and only if at the end of the Strategic Initiative you are happy with what you get.

The result of a Strategic Initiative is delivered based on ongoing negotiation throughout the life of the initiative.

Successful negotiations about the direction and the result of the Strategic Initiative in order to meet the demands of the stakeholders require that the negotiation stakeholders know where their currently executing initiatives are relative to their baseline.

For this purpose, you will need Key Performance Indicators (KPI) related directly to your Strategic Initiatives that can tell you if the initiatives are moving in the right direction with a reasonable performance, where performance is measured relative to the baseline, for example:

- Are we heading at a cost overrun in the end or will we be within budget?
- Do we deliver partial results on time or are we heading to a delay in the end?
- Does the deliverable quality of partial deliveries meet our expectations?
- Are the resources required to progress available on time?

On top of these KPI, we need less tangible performance indicators to tell us:

- Is the solution we are creating still relevant?
- Will unexpected competition prevent us from getting the expected ROI?
- Do our Strategic Initiatives still have the same priority with the sponsor?

I trained the project and program managers of a major Swedish industrialist in project and program management with high weight on communication management and key figure usage from project progress tracking such as Earned Value Analysis Cost Performance and Earned Schedule indexes.

My sponsor and I had developed the training material dedicated to their situation as vendors and implementers of major industrial production structures in a market with strong competitors.

Two years after giving the training, I went back to understand how and with what result they had used what they had learned. This is the answer I received:

> Soren, we were happy to plan and track our programs and projects as you had suggested. We even used the coffee bean to structure the Work Breakdown Structures for agile behavior and tracking, and we delivered reliable tracking information to our sponsors all the time; but our sponsor closed out our project with 32 persons full time occupied from one day to the next without any warning. Even today, we do not understand the precise reason for this action.
>
> How could we have prevented this situation from happening?

There are events that will hit you as manager or participant in a Strategic Initiative that you can do nothing about. Even the best communication and the best negotiation-based relationship with your sponsors cannot prevent this from happening.

In my opinion, this is not a reason for not at least trying to make good tracking and communication. Even though you have nothing to negotiate about you might at least know why.

Negotiation is based on communication. Somebody discovers that deviations from expectations have occurred and this somebody needs to know how to treat this discovery for negotiations to be initiated about how to react.

Ongoing communication of pertinent performance key figures as soon as they are known can ensure that deviations from baseline are discovered early and that these deviations are handled in the best possible way.

Communication means that there is a sender and a receiver of the information and that they both make sure that the other party understands the implication of what is communicated. Somewhere in the Swedish case mentioned previously this did not happen because some pertinent information from Sponsor to Workgroup and even Governance Team was not communicated before it was too late for negotiation and explanation.

In previous chapters, we prepared the tools and methods for how to respond to deviations from baselines and how to execute change. In this chapter, we will cover the information, methods, and tools that can tell us early on when the deviations from baseline are pertinent so that we can initiate plan adaptation and initiate change before it is too late:

- Strategic Initiative KPI
- The Compound Expected Value (CEV) of the Strategic Initiatives with reference to PQA success factors
- Communication

6.2 ESTIMATION AND FORECASTING

PQA is our first foundation for estimation and forecasting when we plan a Strategic Initiative and when we adopt major changes to the Strategic Initiative, that is, when we re-plan the initiative:

- The success factors give us targets of tangible and intangible nature that can be used for negotiation about where to go and for asking questions as to whether a success factor has been achieved.
- The success factors give us an idea about what to implement and why.
- By showing the expected tangible and intangible values of needed solutions and the planned activities to perform their delivery, the success factors provide a base of reference for prioritizing and evaluating the requirements specifications for the solution components to be delivered by the Workgroup Teams.
- The PQA activities with their outline estimated duration and resource usage give us the foundation for more exact estimation and establishment of milestones and baselines against which we can measure progress.
- The milestones tell us the expected delivery date of solution components so that we can have requirements, test scenario, and people ready for simulated and final Accept-Testing.
- Other baseline elements such as the critical path of Workgroup tasks can tell us about the probability of delivery of the final result on planned time.

The forecasting of the product quality is the requirements specification. We measure the quality of delivered solution components against the requirements specification that has been broken down into very specific and tangible test cases and expected test results.

The breakdown of the requirements specification in use cases and workflows prepares for an agile delivery process with agile planning and tracking.

6.2.1 Monte Carlo Simulated Forecasting

Forecasts of activity duration and cost are random. Activity cost is most often measured in number of person-hours worked multiplied by an arbitrary person-hour cost per hour. The activity duration at least in the beginning of a Strategic Initiative is close to the wishful thinking of the sponsor because no one can prove this thinking wrong unless it is completely unrealistic.

The Sponsor's early forecast is probably based on studies of other similar initiatives if they exist, or on opinions of subject matter experts, or on available funding opportunities. Not many managers will protest against this Sponsor forecast before they have gathered some experience from their own initiative execution.

Strategic Initiative Sponsors and Governance Teams know quite well that the initial forecasts are random with very high uncertainty, but once the Workgroup managers have launched an estimate the randomness is most often forgotten and the poor Workgroup manager is punished or blamed for cost and time overruns or might even get a bonus for doing better than forecasted. Punishment, blame, or bonuses are not deserved here as the results achieved are based on pure luck or lack of it.

Forecasts are better used as information for negotiation. It is possible to establish a realistic foundation for negotiation if you break the activity cost and duration forecasts down into task-based forecasts, where the people working on the task have a more realistic idea about how much time and money they need to deliver what is required from them.

Furthermore, if you add realistic information about dependencies between tasks and maximum availability of resources per period, you get an opportunity to use Monte Carlo simulation to tell you:

- The probability of different levels of total activity cost
- The probability of different activity durations

You can find easy to use free or professional tools available for Monte Carlo simulation that is fully integrated with Microsoft Project and Excel.

The Excel spreadsheet with your estimates per task will draw a probability distribution of your random activity duration and cost after simulation of your project plan. The tools allow the forecasting of three cost and duration estimation figures to be entered into the Microsoft Project task information for this simulation to take place:

- Most pessimistic, say in only 1 percent of cases will duration and cost exceed this value.
- Most likely, what we would have committed to if asked for only one figure.
- Most optimistic, say in only 1 percent of cases can we do it so fast and so cheap.

The Monte Carlo tool will simulate between 100 and 1000 plan-cases and tell you the probability of different cost and duration estimates. The number of plan-cases to be simulated is decided by you. Increasing the number of simulations above a certain number will not yield better results. The more advanced tools will tell you what the optimal number of simulations is for your plan. Besides the simple triangular probability distribution, some tools allow you to use more advanced statistical probability distributions such as the normal distribution.

It is now up to you as Sponsor, Governance Team member, or Workgroup Manager to decide on which probability level you want to plan the work:

- If you are optimistic or you want to stress the Workgroup or you hope to convince an external vendor to give you a favorable offer, you will go for an estimate, where you have only 50 percent probability to succeed.
- If you are more realistic, you will go for an 80 percent chance of success and put less stress on the Workgroup to get a better result quality.

Now you can negotiate and both parties can go to work knowing the cost and duration risk they are facing.

Monte Carlo results, shown in Figure 6.1, look like this based on the cumulative probability of being below a given cost and duration figure.

I like Monte Carlo simulation because the results are easy to interpret and therefore provide a basis for negotiations, where you can go back to the three-point estimates on each task and adapt these to new knowledge and experience as you move forward with your Strategic Initiative.

If now the Workgroup Team and its manager over-perform on agreed targets, it is still lucky, but in this case, a team bonus will be good for its

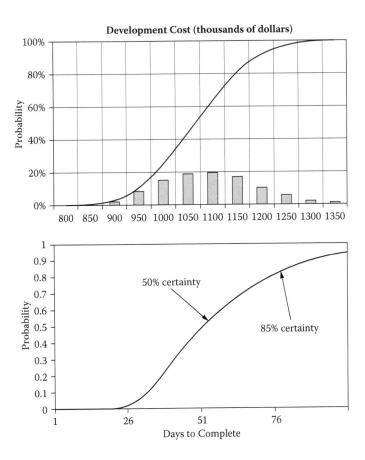

FIGURE 6.1
Monte Carlo accumulated probability of cost and duration as made with software from
e.g. www.palisade.com/Risk.

motivation to keep the performance up and participate in future negotiation and decision making.

6.3 STRATEGIC INITIATIVE KEY
PERFORMANCE INDICATORS

Once initiated, the progress of the Strategic Initiatives is tracked in order
to be able to adapt to new knowledge and other changed conditions.

Over time the risk profiles change and with those the probability of
opportunities and threats.

The tracking is based on KPI that can provide information about:

- Organizational condition changes
- Solution condition changes
- Process condition changes

Some KPI are specific for each condition type, while others look at cross-initiative environment indicators that indicate if duration or cost is under control for the Strategic Initiative.

You will also need KPI that can tell you whether your Strategic Initiatives perform as they should from a business perspective, that is, looking at all opportunities and threats known at a given point in time:

- Is what we are doing still attractive?
- Do we need to change the strategy and re-establish more valuable Strategic Initiatives?

A KPI that can help with Strategic Initiative tracking from a business point of view is the Compound Expected Value (CEV). Calculation of CEV per Strategic Initiative makes it possible to compare the initiatives mutually and to evaluate them in the context of overall business strategy performance.

6.3.1 Classic Strategic Initiative KPI

Well-known KPI comprise:

- Cost Performance Index from Earned Value Analysis that allows you to measure if you are on track cost-wise and can give you an estimated cost at completion.
- Time-based Schedule Performance Index [SPI(t)] from Earned Value Analysis with the difference from classic cost-based SPI being that you measure duration. SPI(t) allows you to estimate the final delivery date, which the cost-based SPI does not allow.
- Critical Path Method (CPM) float calculation based solely on task duration and task dependency can show you if you are early or late by task or in total.
- Baseline Variances (activity delays, missing resources, missed milestones).

- Outstanding issues based on your list of issues with reference to issue documentation and indication of status and expected time to closure.
- Quality control results (error log) from SAT and Acceptance Test that shows an objective status of the quality of solution components to be delivered.

You can look up how these statistics are calculated in project management textbooks or on the web. In order to calculate these KPI, you need information systems and procedures that ensure the necessary discipline from the planning data and result reporting Workgroup Teams.

An example of a quality management COTS is HP Quality Manager.

An example of project management COTS is Microsoft Project.

These applications are COTS applications just like any other COTS application, so they require setup, support, and trained users to give you benefits from planning and tracking. I would never try to use such systems without Project Office support, facilitation, and coaching.

The Strategy Governance Teams can work together with the Project Office if it exists to establish Information Systems in support of planning and tracking of the Strategic Initiative Workgroup activity. If an Information System-based foundation for planning and tracking is not established, the Workgroup Managers and the Governance Teams get the tracking information too late about problems and incidents to respond to these in time because:

- The Workgroup Teams do not discover that they are late until deadlines have passed without delivery of the expected solution in the right quality and until milestones have not been met.
- The Workgroups only report progress when asked to do it in the form of "we're fine and we'll deliver in time" even though more reliable Estimated Work-hours To Complete would tell you and them that this is not possible.
- Even though the Workgroup Team knows that it is struggling to make it on time and cost, the members are too proud and optimistic to admit it.
- The busy Workgroup Team does not discover its dependency on the performance of other Workgroups before needed resources or deliverables do not turn up on time.

- The Workgroup thinks that it can deliver the expected quality on time by working faster or by adding resources although all known experience shows that this is never the case.

The Project Office supports the usage of the required Information Systems to ensure that the Workgroup Managers register the plans from PQA and estimation activity correctly and so that the Workgroup participants can report:

- How many person-hours they need to finish their work.
- How much calendar time they need to complete the work.
- If and when they are available to do the work.
- Expected unavailability not planned.

The Workgroup Manager will approve this information and further add information about dependencies between tasks.

Based on this information, the Information System can calculate delays and cost overruns with key figures such as Cost Performance Index or duration variances on the critical path of the Strategic Initiative, where any delay has an impact on the final duration of the initiative.

On a regular basis, all active Workgroup Managers get together with the managers from organizations that provide the required resources to coordinate that these resources become available as needed or are released for other purposes if they are no longer needed in an already agreed period.

All of this tracking and ongoing re-planning work can function well if supported by a Project Office.

I have never seen project tracking and timely strategy re-planning work well without the support of a Project Office.

The Workgroup activity results in delivery of solution components. Another way to track the performance of this work is by measuring how fast error-free solution components are delivered. This is the agile tracking method.

6.3.2 Agile KPI

When working agile with the development and implementation of working solution elements, you are supposedly not interested in all the nitty-gritty planning and coordination presented previously.

You "simply" get the right people together, provide them with all required material and resources, and wait for them to send out white smoke when a solution has been produced, tested, and approved for production.

Your agile work breakdown structure consists only of use cases or autonomous solution components closed out with passing SAT. The solution components to be produced are your burn down list that you make shorter and shorter by delivering result components.

You control the delivered solution component quality by SAT testing and an outstanding error and issue list that allows you to accept changing conditions and requirements as you get wiser.

The solution that passes SAT is what the stakeholders want, so you have no quality issues once the solution component has been delivered.

In theory, you produce solution components at a fixed rate of speed (velocity) that based on your burn down list allows you to calculate the outstanding duration of the work.

By having all needed resources available all the time, you do not need to do resource planning more than once, when the resources for the agile delivery process have been allocated to the Workgroup.

When planning Strategic Initiatives, you should look for opportunities to work agile. When you manage in an agile way, you can concentrate all your effort on the design of solution architectures with use cases and on building the use case-based solution elements fast and with good quality based on SAT.

To establish and govern the agile opportunity will demand all your project management skills.

Before you get to the opportunity to go agile, you still need to do classic planning and arrange for classic performance indicators to be produced in order to manage all the tasks that are needed to cater to the agile Workgroups:

- Procurement of resources
- Procurements of COTS
- Establishment of development environment
- Establishment of SAT scenarios
- PQA preparation and conduct
- Preparation of the basic requirements specification

6.3.3 Solution Quality Evaluation

Once you have isolated the agile teams, you get the best possible foundation for evaluation of solution quality because solution components passing SAT by definition are accepted by the stakeholders.

These solution elements do not have to be exactly what is written in the requirements specification as long as the stakeholders represented by the Accept-Testers accept them. Quite often, the solution is better than required, and if this happens without increasing the cost such as on fixed price contracts, everybody is happy.

In the private bank swap-case, each Workgroup consisting of developers, IT supporters, and end-user SAT and Final Accept-Test testers was an agile group. That their result was finished and ready for use was underpinned by the fact that this result released invoicing and immediate payment to the involved external solution providers.

Once the Workpackage producing Workgroups had been established, the Governance Team had no interest in intermediate results, only in fulfilling the changing demand for resources and material that was requested by the Workgroups ongoing as they acquired experience with setting up the needed solution components and the COTS.

As not all use cases were equally complicated, the Workgroup Managers had some classical planning to do in order to provide the Governance Team with progress information and expected delivery dates, but there was no control needed from the Governance Team side, only support to ensure "no excuse for failure."

Based on the agile strategy and the clear rules of solution acceptance, there were practically no outstanding errors and issues except for the ones that originated from an exceptionally bad quality of the underlying COTS. The error and issue documentation was of great value during negotiations with the COTS vendor.

When you are not working in an agile way based on technically error-free delivered use case-based solution components, the evaluation of the solution quality is more complicated:

- You have no guarantee that the delivered solution components are free from errors that have nothing to do with your business needs. This means that when you find errors or when the solution component simply does not work, you and quite often the delivering Workgroup do not have a clue as to what is wrong. Error correction will demand profound analysis before the reason for the error is found and a correction can be implemented.
- You have to develop test cases that concern COTS functionality that is not relevant to your business processes, but that have to be set up for the COTS system to work.

- If your solution component communicates with already installed information systems, it is quite possible that an error in the original system that had no influence on this system on its own will prevent your new system from communication. Errors like this can be difficult to correct because the reason is hidden somewhere that you cannot know about before you trace the system functionality until the error occurs in a controlled manner.
- The COTS vendor only guarantees that the COTS system works as documented in the user guide and the installation guide, but it is up to you to find the errors in COTS. Finding errors in COTS is not easy because you most often think that you have made a setup error, and it can take a long time before you can prove that the reason for the error is COTS and not you.
- It is difficult to know when you have finished testing because corrections can lead to new unforeseen and unknown errors that might be more serious than you thought could be possible.

While testing a new "Know Your Client" COTS application in a major bank, the COTS worked well on its own, but once we tried to load client data from the central banking application the COTS hung up and the users refused to accept the solution. In this case, the reason for the error was the message queue application that had been especially adapted to the central banking system data types, which were not supported, by the "Know Your Client" application.

The message queue system was no longer supported and the new owner of this demanded an unacceptable fee to re-establish support without guarantee that this support could solve the problem.

Fortunately enough, the original developers had started their own business and they were willing to fix the problem if we had access to the escrow source, which we had.

The problem was small when we found the reason and the skilled people to fix it. However, it took 6 months to get this far with multiple new error corrections from the vendors of COTS and the "Know Your Client" application for which the bank had to pay. Furthermore, it resulted in a lot of testing of new versions that did not work, taking time from busy bank employees.

In the same bank we were installing a major new release of a COTS-based Client Relationship Management (CRM) application introducing MiFID compliant procedures and new reporting.

This new COTS version installation was concerned with new functionality and reports delivered from the COTS vendor based on a simple release document that explained implemented changes since the last release:

- There was no requirements specification from the bank's side.
- There was no test model or test cases.
- The test of the new release was performed by a consultant from the COTS vendor and the IT support person based on pure intuition.
- The only issue documentation was e-mails sent back to the COTS vendor with errors for correction.
- There was no issue list produced to track the issue status.
- The only response to the errors and issues was new release documents and new bug fixes.

This iterative Accept-Testing dragged on and on and only closed out because the bank IT management and future user test management declared the new release ready for delivery to be Accept-Tested by the user test group.

The user test group had a set of standard test cases that it applied on the new release and that immediately showed production blocking errors.

Furthermore, the CRM COTS interfaced the central banking solution that worked with completely different periods and data types resulting in other production and usage blocking errors that had not been tested by the CRM COTS vendor and the internal COTS support person.

Finally, production and usage was established by force with a delay of 8 months because of immediate business need. Usage required daily manual corrections in both systems until both systems were finally swapped out.

The lessons learned are as follows:

- You cannot implement solutions without requirements specification.
- You cannot test a solution without test model and test cases that are complete. If you are not agile, you need many more test cases on the user side to be complete:
  - The test cases are complete with respect to the requirements specification.
  - The test cases are complete with respect to the release document.
- You cannot test information system functionality properly without testing business process functionality and COTS functionality (release documented) concurrently.
- The people testing, reporting errors, correcting errors, and finally delivering into production must be competent.

6.4 THE COMPOUND EXPECTED VALUE

The Strategic Initiative Sponsor and the Strategy Governance Teams require KPI to measure the progress of Strategic Initiative benefit delivery. It is more complicated to establish such KPI than to establish the ones that relate to project progress and solution quality tracking. Most often benefit delivery KPI are confined to indexes or absolute values showing their development over the period of the Strategic Initiative implementation:

- ROI
- Number of employees
- Profit growth
- Corporate value expressed in share price and stock exchange value

An example of corporate KPI is an extract from a corporate annual report with figures from 2011 and 2012 shown in Figure 6.2.

Strategic Initiative management has a problem with these figures because they only tell us where the corporation is heading based on historic figures. We cannot see if the currently executing Strategic Initiatives are performing well.

We can only see how Strategic Initiatives have succeeded in the past, and even here, we do not get an evaluation of the progress of each initiative.

We cannot use these KPI to give us an indication about in which direction to change our strategy; they can only and normally much too late tell us that a change of direction is needed if we want better key figures next year.

What we can get from historical KPI is at best a benchmark:

- We get an impression of the magnitude of the organization with which we work.
- We can see what management and shareholders regard as important.
- If we read the annual report, we can get a lot of detailed information about markets, market share, products, services, organization structure, geographic location, etc.
- We might be able to see what sort of information is important as arguments for initiating and implementing a Strategic Initiative:
  - Entry into a new market with high growth and profit expectations
  - Improved employee efficiency
  - Improved logistics management
- Higher product quality

Key figures, EUR million	2011	2012
Net sales	6,646	7,504
Services net sales	2,871	3,174
Services, % of net sales	45	44
Earnings before interest, tax and amortization (EBITA) and non-recurring items	628.5	684.3
% of net sales	9.5	9.1
Operating profit	571.8	598.5
% of net sales	8.6	8.0
Profit before taxes	507	550
% of net sales	7.6	7.3
Profit	358	372
% of net sales	5.4	5.0
Procurements*	4,319	5,020
Research and development expenditure (including IPR expenses)	137	139
Research and development personnel	852	831
Priority applications, pcs*	180	215
Invention disclosures, pcs*	649	679
Gross capital expenditure (excluding business acquisitions)	166	156
Business acquisitions, net of cash acquired	15	5
Earnings per share, EUR	2.38	2.49
Dividend per share, EUR **	1.70	1.85
Balance sheet total	6,618	6,642
Return on capital employed (ROCE), %	18.4	19.6
Return on equity (ROE), %	17.8	17.5
Equity to assets ratio, %	39.8	40.5
Gearing, %	12.2	14.2
Free cash flow	375	257
Orders received	7,961	6,865
Services orders received	3,100	3,264
Order backlog, December 31	5,310	4,515
Personnel, December 31	30,324	30,212

FIGURE 6.2
Corporate KPI examples.

However, how do we measure that our Strategic Initiatives will deliver the expected benefits from these initiatives?

Why and how do you assess that the underlying Business Case is still valid?

Does the original idea hold or do we have to improve or change strategy completely because the corporate situation has changed?

How do we know that the required benefits have been achieved and a new strategy might be needed?

We need a system to give early warning on strategy level, not only on Strategic Initiative level, a system and KPI that can tell the Strategy Governance Team why the Strategic Initiatives can be improved and by what means.

On this background, it is suggested to use the more strategic CEV to complement the classic Strategic Initiative KPIs.

The CEV of the strategy is calculated over the full lifecycle of the strategy:

$$CEV = \text{Compound expected benefits (CEB)}$$
$$-\text{Compound expected costs (CEC)}$$

CEB = Σ P{opportunity}*Impact (sum over all opportunity events)

CEC = fixed cost + Σ P{threat}*Impact (sum over all threat events).

The fixed cost element is the initial investment in the strategy implementation in order to ensure that the agreed quality of the strategy is attained (P{fixed cost} = 1).

All other costs and benefits depend on the incurred risk, where the most important opportunity is that the strategy delivers the agreed and expected result; and the most important threat is that the strategy for some reason fails to deliver the expected benefits.

CEV is based on risk and therefore looking into the future.

The probability and impact factors applied by the Strategy Governance Team using the CEV to evaluate the corporate strategy and the Strategic Initiatives contribute to explaining the current view on the future opportunities and threats based on what has been experienced with the Strategic Initiatives and other corporate events until the evaluation date.

The initial events can come from the success factors and risks documented under PQA (Figure 6.3).

CEV offers several benefits:

- It can be used to compare Strategic Initiatives of different types.
- It can be accumulated for any set of Strategic Initiatives with adaptation to the risks changed by establishing this set (the synergy effect of the set of Strategic Initiatives)
- It can be used to compare individual Strategic Initiatives for choosing the most appropriate ones for implementation.
- CEV calculation gives full traceability of the calculated value in terms of the probabilities assigned and the impact evaluations used over time.

Events (e.g., success factors)	Impact (€)	Probability (0–1)	CEV (€)
Planning culture with respect for others planning	500.000	.8	400000
Provides insight for others into my projects	500000	.5	250000
The invoice foundation appears significantly faster	100000	.9	90000
Solution is delayed more than 6 months	−400000	.8	−320000
The cost overrun is bigger than €1 mill.	−1000000	.2	−200000
Strategic Initiative Total CEV			**220000**

FIGURE 6.3
CEV calculation example.

- CEV will change throughout the lifecycle of the Strategic Initiative under the influence of progressive elaboration, for example, new experience, new knowledge, improved decisions, changed conditions, and new or obsolete risk.
- The CEV value is a strong indicator of whether the strategy is in or out of control, if the organization establishes relevant thresholds for the CEV value, that is, acceptable minimum and maximum values.

6.5 COMMUNICATION

We have already used several types of communication in the previous chapters:

- PQA
- Meetings for initial scope definition
- Presentations for PQA and IRS participants
- IRS Interviews
- OLA-based IRS consolidation
- 9:00 meetings for progress check and fast reaction to issues
- Change request tracking from Workgroup to Strategy Governance Team
- Status meetings to announce major changes in direction and objectives

All of this communication is planned for and supported by standardized documentation and presentation formats that can be adapted to specific situations or improved over time as experience is obtained.

The most important communication tool is probably PQA because it builds on a profound stakeholder analysis and lets us understand the stakeholder attitudes and wishes. Much better than with PQA, we cannot initiate our foundation for communication management.

Furthermore, the rules of PQA conduct pursue high quality of the information that is communicated in order to ensure that the communication becomes SMART. The PQA process ensures mutual respect, which contributes to ensuring that the PQA-based communication becomes:

- Courteous (people problems are not treated under PQA)
- Concise (concrete, pertinent, precise)
- True
- Coherent
- Complete
- Credible
- Valid
- Creative

Unfortunately, we cannot do PQA all the time, so we also need other tools to ensure that the strong team feeling and mutual respect created by PQA stay intact until the Strategic Initiatives close out.

We need to communicate in order to keep all stakeholders happy all the time or at least to make sure that the stakeholders understand why they have a reason to be unhappy.

If a stakeholder does not understand what is going on, we can be sure that this stakeholder will be unhappy and suspicious at some point in time, especially if the stakeholder has a reason to fear that the Strategic Initiative results will have a negative effect on the personal or professional life of this stakeholder and the stakeholder's organization.

Stakeholder unhappiness based on not knowing why the Strategic Initiatives are implemented and how they will influence their personal or work life leads to conflicts and discussions that you want to avoid. Such conflicts and discussions can be avoided with targeted communication and negotiation based on valid and concise information about the solution, the organization, and the process of the Strategic Initiative and how the initiative is planned to have influence on the stakeholders, as well as where this influence is negative.

You might not be able to establish agreement with all stakeholders, but you can certainly establish visibility and get to understand why someone might try to block the progress of the Strategic Initiative.

6.5.1 Workgroup Manager Communication Management

As a Workgroup Manager, you are also a member of a Process Governance Team. In your role as Workgroup Manager, you manage the communication in the Workgroup and between the Workgroup and the Governance Team. You also manage the communication between the Workgroup and the internal and external stakeholders whether these are other Workgroups, external COTS vendors and solution providers, or internal departments that provide resources or expect delivery of solution components.

In your role as Governance Team member, you represent the Workgroup in the Governance Team and you communicate the Workgroup status, for example, new resource and material requirements, KPI, and the issue and error situation with the Governance Team.

The objective of your communication is to ensure that "no excuse for failure" is obtained through negotiation:

- With the Governance Teams, you want to obtain whatever scope changes are needed with respect to funding, time, and solution quality. This negotiation is based on what you know from dialogues with team members, the issue and error documentation, and the Workgroup KPI. Scope changes and new baselines are initiated and negotiated by you in the Governance Team that can approve changes on its own or after further negotiation on a Strategy Governance Team level.
- Within the Workgroup, you ensure that development and implementation teams work together or at least communicate in an agile way.
- You ensure ongoing "no excuse for failure" conditions.
- You listen to Workgroup resources in order to ensure efficient solution delivery and fast and pertinent adaptation of the work and the solution components and sometimes the resources based on the daily work experience and approved changes from the Governance Team.
- You call the Workgroup Team together to inform them about needed changes and you negotiate how to implement these changes with the

Workgroup Team members. This negotiation about how to implement change can sometimes take place under PQA-like conditions, especially if changes to the Workgroup Team are involved.

- On a daily basis, you meet the active Workgroup Team members in person-to-person dialogues and in 9:00 meetings simply to keep the team spirit high, but also to discuss issues and problems of both personal and work natures.
- You make sure that you have a reliable and competent deputy manager to replace you in case of your absence.
- With internal and external stakeholders, you negotiate the resource needs and ensure that these resources are informed and motivated to work in or be used in your Workgroup team during development or implementation and SAT. The communication here is negotiation based on knowledge about not only your own Workgroup needs, but also insight into the status and needs of other Workgroups demanding the same resources as your Workgroup.

When I see Workgroup Managers personally refining Gantt charts and other planning instruments to a detailed level that is not required for communication, negotiation, and establishment of KPI instead of communicating and negotiating, I know that something is wrong.

When I see Workgroup Managers developing requirements specifications instead of communicating and negotiating, I know there is a high probability that the Strategic Initiative will fail. The Workgroup Manager will defend this situation with a lack of qualified resources in time and limited access to funding, but these are bad excuses. If the manager spends the time on planning, communication, and negotiation, the funding and the resources will be available in time because all stakeholders have agreed to the arguments for this fact. If you just do the work, there is no more room for negotiation and everybody loses time and, even worse, they lose motivation.

Workgroup manager communication, negotiation, and the preparation of this is a full time job. The manager job is to ensure the timely availability of qualified and competent resources that can perform development, IRS, OLA, implementation, testing, training, etc. much better than the Workgroup Manager can because they get the time demanded to perform these tasks. The manager is the guardian of "no excuse for failure."

IRS and operation of the Planning Information System can be left with other resources such as Facilitators and Coaches from the Project Office and a Program Office:

- Once the project plan has been estimated, the Project Office can make sure it is registered in the Planning Information System for correct tracking and reporting.
- The Project Office can ensure and control that Workgroup team members add task status information in the form of person-hours worked on tasks and work effort estimates in the form of person-hours to complete the task. As Workgroup Manager, you are still responsible for reliable time and cost to completion forecasts because this is a result of negotiation.
- Project Office can ensure that the Workgroup Manager gets the pertinent information for plan adaptation and KPI generation as required by this manager.
- Professional business analysts and facilitators from the Program Office perform IRS and OLA best; seldom do the Workgroup Managers.

Management is about communication and negotiation in the context of planning and tracking of the Strategic Initiatives.

Planning and tracking systems and techniques are communication and negotiation tools that can be driven by others once the Manager has laid out the foundation with PQA and other communication scenarios.

As Workgroup Manager, you establish a situation with reliable planning and tracking information supported by the Project Office and the Program Office that allows you to understand where you are and to think ahead. You need detailed progress information reports that show organizational performance (productivity), resource availability, cost control information, procurement information, and much more that let you adapt quickly to situations of problems and new risk before major changes are needed.

6.5.2 Governance Team Communication

Governance Teams and Sponsors need information that allows them to act and react fast on unexpected results, events, conditions, risk triggers (events or conditions that imply a higher probability of a known risk), and new risk.

It is quite popular to provide this information in the form of corporate scoreboards with a variety of KPI. As debated previously, scoreboards are worth nothing if the figures do not compare to expected values, that is, the values established by PQA or some other agreed benchmark.

When we talk Strategic Initiatives, the corporate KPI such as shown above are on a too high level to be useful for strategy governance. One size or type of information does not fit all stakeholders:

- Leaders and other Strategic Initiative sponsors need information that shows whether the delivery of the required results is on a good track. This information is not only in report form, but also is presented and discussed in a dialogue between the involved stakeholders in order to evaluate the real progress and to negotiate about what can be done to improve the situation.
- Valid interpretation of pertinent Strategy Initiative scoreboard statistics such as CPI, SPI(t), error and issue lists, change requests, Gantt charts with tracking, and critical path float might require involvement of competent subject matter experts (SME).
- It is the Workgroup manager's interpretation of the Strategic Initiative progress information that is the most valuable information required by the Governance Teams and the strategy Sponsors, not the progress information itself.

The Workgroup Manager interpretation can be presented as a PowerPoint presentation or in more formal status reports accompanied by change requests as needed, but none of these reporting elements can stand alone; they always need a personal presentation by a competent manager.

6.5.3 Project Office Communication Support

The acquisition and presentation of valid and pertinent tracking information in organizations with many initiatives demanding the same scarce resources is a complicated task that cannot be left to a single Workgroup Manager.

Each Workgroup Manager will report and validate their part of the Workgroup KPI foundation, but the consolidation of this information is more technical than management oriented.

The Project Office is needed to produce and ensure the technical validity and completeness of the tracking information across business functions, departments, and Strategic Initiative Workgroup Teams.

Information validated and completeness checked by the Project Office can be used by Workgroup Managers and other internal and external stakeholder management to produce pertinent status reports and PowerPoint presentations to the Governance Teams and the Sponsors on a regular basis.

The responsibilities of the Project Office are secretarial in this context.

Contrary to many beliefs, the Project Office does not develop or implement corporate standards for Strategy Initiative management and execution. Neither does the Project Office establish the corporate Strategic Initiatives. These tasks are handled by corporate leadership and management such as shown in previous chapters.

Once the standards are documented and agreed to on a corporate level and sometimes on a Strategic Initiative level, the Project Office can facilitate and coach the Workgroup Managers by using the standards to verify that the reporting and information produced by Workgroup Management has a quality that allows it to be used for Governance Team and Sponsor reporting, presentation, and decision making.

The Project Office does not interpret reports and information, it only makes sure that the information is delivered in a timely manner on a form that allows the reporting to be produced, distributed, and communicated.

The role and responsibility of the Project Office should not be underestimated. In organizations with many ongoing business operations and Strategic Initiatives, the standards-based collection, validation, and distribution of basic and consolidated planning and tracking information is a complex task.

6.5.4 Program Office Governance Team Communication Support

The Governance Teams do not have a permanent presence such as an ordinary business organization. The Governance Teams meet periodically or they are assembled for solving critical issues and to respond to demands for change from Workgroup Managers or Governance Teams on a lower level. It is therefore essential that the Governance Team be supported by a professional secretarial function—the Program Office.

A Program Office can be specific to a Strategic Initiative or it can have the responsibility to support all Strategic Initiatives.

The Program Office can be interpreted as the executive organization of the Strategy Governance Team.

The Program Office coordinates the scope of the Governance Team's information needs with the Governance Teams and the Workgroup Managers supported by the Project Office.

The Program Office experts lobby with external stakeholders in order to understand their needs and expectations. In this context, the Program Office produces reports and requirements of legal and other compliance nature that are important for the decision making events such as PQA on the Strategy Governance level.

The Program Office facilitates and coaches PQA and Risk Management workshops on behalf of the Governance Teams that it supports based on the results of their particular investigations of external stakeholder demands.

The Program Office can perform IRS if it has been established to have access to the needed skill and competence.

6.6 LESSONS LEARNED

While projects are relatively easy to manage because they produce predictable and tangible results with known resources and well-defined baselines, once they have been planned and are executing the program part of Strategic Initiatives is more difficult to manage because the benefit target is much less tangible.

The outcome of Strategic Initiatives is random and for both forecasting and tracking we need to apply statistical methods in order to understand the magnitude of variance with which we are faced.

In the Strategic Initiative world of random outcomes, agile planning and tracking of our activity allows an easier to understand follow-up on progress. By measuring progress only by user and management accepted solution components delivered, installed, and ready for production, we know at least for sure what we have and we have a steadily improving foundation for estimation based on the gained experience.

Whenever it becomes evident that the original targets cannot be met or that the original targets no longer are valid, it is time for change of strategy. Changes and adjustments to solution scope, organization, and process make the Strategic Initiatives fit the reality that becomes visible only as the Strategic Initiatives progress and meet the obstacles that we did not expect and for which we did not plan.

Strategy governance is negotiation.

Negotiations without one dominating party is a first step on the way to obtaining win-win solutions and synergy effect, while domination from one party has a demotivating effect that most often touches all involved stakeholders because the dominating party will have doubts about the effect of what was dictated in light of the demotivation of the dominated party.

Negotiation is based on communication. Somebody discover that deviations from expectations have occurred and this somebody needs to know how to treat this discovery for negotiations to be initiated about how to react.

Communication means that there is a sender and a receiver of the information and that they both make sure that the other party understands the implication of what is communicated.

PQA is our first foundation for estimation and forecasting.

Strategic Initiative sponsors and governance teams know quite well that the initial forecasts are random with very high uncertainty, but once the Workgroup Managers have launched an estimate the randomness is most often forgotten and the poor Workgroup Manager is punished or blamed for cost and time overruns.

Forecasts are better used as information for negotiation. It is possible to establish a realistic foundation for negotiation if you break the activity cost and duration forecasts down into task-based forecasts, where the people working on the task have a more realistic idea about how long they need to deliver what is required of them.

When planning Strategic Initiatives you should look for opportunities to work agile. When you manage in an agile way, you can concentrate all your effort on the design of solution architectures with use cases and on building the use case-based solution elements fast and with good quality based on SAT.

Once you have established the agile teams, you get the best possible foundation for evaluation of solution quality. The solution components that pass SAT are, by definition, accepted by the stakeholders.

When you are not working in an agile way based on technically error-free delivered use case-based solution components, the evaluation of the solution quality is more complicated.

We need a tool to give early warning on strategy level, not only on Strategic Initiative level, a tool to deliver KPI that can tell the Strategy Governance Team why the Strategic Initiatives can be improved and by what means. Establishment and calculation of the Compound Expected Value is such a tool.

We need to communicate in order to keep all stakeholders happy all the time or at least to make sure that the stakeholders understand why they have a reason to be unhappy.

Stakeholder unhappiness based on not knowing why the Strategic Initiatives are implemented and how they will influence their personal or work life leads to conflicts and discussions that you want to avoid.

You might not be able to establish agreement with all stakeholders, but you can certainly establish visibility and get to understand why someone might try to block the progress of the Strategic Initiative.

The objective of your communication is to ensure that "no excuse for failure" is obtained through negotiation.

Workgroup Manager communication, negotiation, and the preparation of this is a full time job.

When I see Workgroup Managers developing requirements specifications instead of communicating and negotiating, I know that the Strategic Initiative will probably fail.

IRS and operation of the Planning Information System can be left with other resources such as facilitators and coaches from the Project Office and a Program Office.

Management is about communication and negotiation in the context of planning and tracking of the Strategic Initiatives.

The Project Office is needed to produce and ensure the technical validity and completeness of the tracking information across business functions, departments, and Strategic Initiative Workgroup Teams.

The Program Office can be interpreted as the executive organization of the Strategy Governance Team.

7

Agile Strategy Management Recap

The corporate strategy is the reason behind the structure and the behavior of the corporation. The strategy comprises the corporate:

- Vision
- Mission
- Objectives

The corporate strategy is ongoing adapted to market condition by Strategic Initiatives that establish the WHY, the WHAT, the WHEN, the HOW, and the WHO concerned with sustaining, changing, and improving business procedures and infrastructure in support of the corporate strategy.

Strategic Initiative management uses the agile principles for handling teams, change, and continuously improved business quality (Figure 7.1).

A Strategic Initiative can be concerned with reaching many different objectives concerned with different business situations.

The corporate leaders provide the initial set of objectives in response to an identified need for change of business conditions. These objectives are an indication of what kind of business and stakeholders must be involved in the establishment and governance of Strategic Initiatives.

7.1 STAKEHOLDER INVOLVEMENT

In order to establish a strategically aligned solution to be delivered through Strategic Initiatives, you deal with a multitude of stakeholders representing all the roles directly involved with development, implementation, quality management, usage, governance, etc. of the required solution as

FIGURE 7.1
Agile strategy management process cycle.

well as the not always visible stakeholders that potentially benefit or suffer from the Strategic Initiative and its solutions.

When establishing a Strategic Initiative you make a serious effort to get to know all the stakeholders that are concerned, that is, to have a dialogue with key persons and organizations that potentially could benefit or suffer from it. This is especially true for the less visible and less obvious stakeholders such as unions, politicians, government, legal bodies, and potential competitive businesses and partners.

In several cases, leaving out potential stakeholders has led to the complete failure of the strategic effort.

7.2 AGILE TEAM BUILDING FOR STRATEGIC INITIATIVES

Agile team building is concerned with the establishment of the best possible organization to perform the Strategic Initiative and adapt it to changing situations and events:

- Selection of people to become key-stakeholders in the Strategic Initiative
- Establishment of the teams of people and the roles and responsibilities of the people in the teams

- Definition of the roles and responsibilities of the teams to perform the tasks required during the lifecycle of the Strategic Initiative
- Establishment of the physical and technological environment within which the chosen people can act and communicate in an optimal way
- Establishment of standards to be used for processes, documentation, and deliverables in order to manage the quality of work, deliverables, and final solution delivered by the teams

Team building is a way to generate synergy, that is, the teams are organized in such a way that the performance of any team is higher than the performance measured as the sum of the single team member's performance.

Only if the chosen stakeholders feel that they contribute to something valuable can you keep them motivated. We keep this feeling alive by involving them in Strategic Initiative processes where they can and will contribute positively and visibly to the result.

7.2.1 Sponsor and Coach/Facilitator Roles

The Sponsor has knowledge about who the key-stakeholders might be and the Coach/Facilitator has knowledge and experience about how stakeholders can be treated and made happy.

7.2.2 Teams for Agile Strategic Initiative Governance and Management

The typical organization constructs involved with Strategic Initiatives from initial establishment to final implementation and governance are:

- A Project Office established in the line organization in support of all projects and programs in the corporation
- A Program Office established as an executive organization representing one or more governance teams
- Decision-making and executing teams established for the development and implementation of Strategic Initiative results under continuously changing conditions and risk.

The different types of teams have one set of capabilities in common (Figure 7.2).

FIGURE 7.2
Team Competences.

The different team constructs with which we have worked under Strategic Initiatives comprise:

- The Strategy Governance Team initiates a Strategic Initiative based on leadership decisions. It comprises the top level of Change Management, the top level PQA Team, and the Change Control Board.
- The Process Governance Team coaches one or more PQA Teams and is the second level of Change Management.
- The PQA Teams lead, manage, and plan activity in order to deliver agreed tangible and measurable results.
- The Workgroup Teams perform production and implementation of agreed solution components. They control the quality of delivered solution components. They report progress in the Project Management Information System that is supported by the Project Office. Problems, Risk Conditions, and Events are reported to the PQA Team that manages the Workgroup Team.

7.2.3 The "No Excuse for Failure" Principle

Your key-stakeholders are probably in high demand in many other activities and therefore they need to be informed of the importance of your Strategic Initiative in order for them to understand what benefits they can obtain from contributing to it.

Facing this stakeholder risk situation, your first response to the risk is to ensure that your Strategic Initiative complies with the "no excuse for failure principle":

- You know why you need the key-stakeholder in your Strategic Initiative and you have a list of arguments that show the value for this key-stakeholder to contribute to your project.

- You know which internal and external activity that will compete for key-stakeholders with your Strategic Initiative and you respect their importance as well.
- Because you are involving people with very different skills, experiences, and competences, you know that conflicting interests are inevitable. You have organizational elements and procedures in place to avoid conflicts becoming personal with a negative impact on the Strategic Initiative progress.
- By using professional coaching and facilitation, you ensure that conflicts only result in lateral thinking (out of the box) and synergy on workshops and during other types of teamwork,
- At any point in time teams and key-stakeholders have access to all pertinently needed and available resources and knowledge constrained only by accepted limits to their availability,
- You plan to ensure that all required resources to be involved in an activity are available and allocated to the activity before the activity is initiated with assignment of these resources,
- You do not initiate an activity if you know that any required resource is not available to be assigned to the activity.

You incur important risk by not complying with the "no excuse for failure" principle:

- Biased strategy focus because important knowledge or competence is left out initially might lead to development and implementation of solutions that do not comply with stakeholder needs—you will lose capital and time.
- Key-stakeholders might lose confidence in the Strategic Initiative because the not involved but required resources raise pertinent critiques of the chosen initiative scope and objectives—you will lose time and key-stakeholders might leave the initiative.
- If the involved resources do not have the competence to reach a conclusion about critical success factors and the way forward to an agreed solution, then the key-stakeholders waste time and lose confidence in (your) management.
- Important processes might be performed with interruptions because of lack of important resources, which leads to waste of time and bad results.
- The initial enthusiasm of the key-stakeholders can disappear very fast if you do not keep them motivated by immediately involving

them in pertinent Strategic Initiative activity, where they get a chance not only to prove their competences, but also to use this competency directly in cooperation with peer stakeholders.

- If the key-stakeholders lose interest in your Strategic Initiative, then the initiative might already have failed.
- If the key-stakeholders get into negative conflict with you or with each other while conducting the Strategic Initiative activity, the initiative is probably already doomed to fail.
- If some resources accuse other resources of failure, it creates stress and negative conflicts that are the reason for delays that could have been avoided by better selection of resources, better team building, and better working conditions.

7.3 STRATEGY PROCESS QUALITY ASSURANCE

In order to succeed with the solution delivery, the teams establish a complete set of plans that with the highest possible probability lead to solutions accepted by the stakeholders.

Planning and plan execution of Strategic Initiatives is not just Project Management, it is to an even higher degree Risk Management:

> The objective of strategic initiatives is to reach FUTURE situations and conditions with high PROBALITY that will provide the IMPACT wanted by the involved stakeholders.

Strategic Initiatives are risk. They can fail or succeed. In order to optimize the chance or probability of success we apply risk management to the Strategic Initiatives. Project Management on its own will not do the job.

Risk Management performed efficiently can allow the teams involved with Strategic Initiatives to build plans that with higher probability achieve the solutions and results (the impact) demanded by the stakeholders.

In this respect, Strategic Initiative PQA is Risk Management and in the work that we have performed Risk Management is PQA.

Please remember that we are always faced with pertinent unknown unknowns and unknown knowns that are ready to surface at any point in time in the future of our Strategic Initiative.

Risk responses are always built into the plan. You respond to risk by adapting your plan to:

- Accommodate the best possible resources
- Utilize the best possible procedures, standards, and techniques
- Adapt the solution to be SMART (Specific, Measurable, Achievable, Realistic, Time bound)
- Ensure satisfactory funding by efficient stakeholder communication

PQA is the method for Strategic Initiative establishment and planning based on intensive teamwork in the PQA Teams with brainstorming that documents the agreed Strategic Initiative for implementation.

PQA is used to ensure the quality of the initial plans, but it is also used to ensure the quality of changed plans, especially in connection with PQA review workshops that are used to respond to risk and to adapt the plans to required changes decided by the Strategy Governance Team.

PQA ensures:

- Identification of the strategy sponsors and other key-stakeholders
- Establishment of the agreed strategy with detailed objectives, Strategic Initiatives, teams and team organization, management, and communication that can ensure the success of the strategy under fast changing conditions and high risk
- Establishment of standards for processes and documentation that can answer the basic questions about:
  - Where we are
  - Where we want to go
  - Why we want to go there
- How we want to go there

The answers are given in terms that can be easily interpreted and agreed to by all involved stakeholders:

- Implementation of the strategy with timely execution of decided strategic initiatives, timely measurement and approval of results and benefits, and efficient change management in support of strategy governance.

7.3.1 Other Strategy Quality Management Tools

PQA cannot ensure the quality of the entire strategy on its own. If you want to know where you are compared to where you want to be while executing a Strategic Initiative, you need procedures and tools other than PQA:

- Project and program tracking is based on a number of performance indicators that can tell you if an activity is delayed, if the project or program is delayed, and quite often if this delay is curable, that is, if changes to the baseline are needed.
- Analysis of information system requirements and solution design is a strong foundation for time and cost estimation. The developers and implementers of the solution can give you reliable feedback about solution quality progress related to usage of resources and funds that can be used for change management, where PQA comes back into the picture.
- Breaking the delivery down into manageable Work Packages based on easily verifiable use cases that are not started, in production, or delivered makes it possible to build Work Breakdown Structures (WBS) for agile solution development and implementation that ensure visible progress and fast and reliable reactions to change requests.

7.4 SOLUTION PROVIDER PROCUREMENT

This is an example of a relatively complex procurement of COTS systems and solution provider services that went well under specific conditions.

Procurement establishes a future required situation, in our case:

- New COTS systems delivered and implemented
- Solution provider service level agreements
- User training
- System operation

In order to succeed with procurement you need to manage the risks involved with this process. Some of the more important risks are:

- If you buy a COTS system before you have a detailed requirements specification, you will probably waste time and money because setup and implementation to fit your (unknown) needs will be a trial and error process until a feasible solution is established with very low probability of success.

- If none of the potentially available COTS systems can contribute to your Information Systems needs without major changes or additions to their functionality, you might get important cost increases and solution delivery delays. In this case, we do not talk about parameterization and setup changes, but about changes not supported by the delivered COTS functionality.
- A chosen COTS vendor always obtains a de facto monopoly once chosen and installed in your IT environment. You will be very weak in negotiations and might incur long lead times and high costs for adaptations in support of your business operations, especially facing business needs that are particular to your business if you do not foresee and include the conditions of these changes in the vendor contract.
- If the COTS vendor goes bankrupt or in other ways ceases to do business, you lose support of your COTS system and you might need to procure another one. This is the background for the escrow clause in the vendor contract. The escrow does not prevent you from losing money and time, but it might help you to protect your business information.
- If the chosen solution providers and the COTS vendors do not have enough competent resources to set up the COTS systems and integrate them into your IT environment, you will encounter big delays in implementation and increased cost that you cannot recover.
- The installation test performed by you and the COTS vendor is made on the COTS vendor's contractual terms—you get what you see and that is all that the COTS vendor will guarantee. However, this does not rule out serious errors seen from your point of view that the vendor looks at in a different way. This can create conflicts, delays, and increased cost.
- If the best COTS solution does not use the same IT infrastructure that you have installed, the acquisition of this COTS system will require that you add to or change your IT infrastructure, which will add costs to not only new technology, but also the knowledge and organization necessary to run the new technology.

7.4.1 Procurement Lessons Learned

- If a project comprises many sub-projects and broad organizational involvement, you establish such a project as a program with a Strategy Governance Team and a competent Change Management Board. In our case, the vendors were invited to participate on the Change Control Board, which avoided all conflict.

- Do not procure anything without a good requirements specification.
- If you are faced with a monopolistic vendor and you have a weak service level agreement (SLA) seen from your point of view, you ensure access to competent knowledge on your side before you negotiate any further conditions with this vendor. If not, you risk losing even in court.
- Trust your partners only if this trust is based on a good contractual foundation (SLA). This is the best way to avoid litigation and complete program failure.
- Internally delivered Work Packages are agreed to in writing with the internal "solution provider" just like the external ones.
- Internal solution provider personnel prove their competence based on pertinent CV data just like the external ones.
- Demand weekly tracking of progress from both internal and external Workgroups based on reliable WBS and estimations of time to complete (cost tracking is not an issue because delivery is on fixed price).
- Use a project management COTS system in support of WBS registration, planning, and tracking.
- Any program and project is risk and has to be governed by risk management (PQA) in order to ensure efficient risk response.
- Risk originates from solution, process, and organization elements, and risk response encompasses management of the quality of these elements.
- Do not be afraid of doing what no one has done before if this is the way you can respond to your risk situation.
- Things are never so urgent that you do not have time to do what has to be done right.
- Make sure that all victories are victories for all involved parties.
- Get the key-stakeholders activated even though they claim that they have no time to offer.

7.5 STRATEGY IMPLEMENTATION

Strategy implementation is doing what has been planned in order to ensure that the Strategic Initiatives deliver the expected benefits to the organization.

The Strategic Initiative Teams know their roles and responsibilities to deliver the expected quality of all agreed results. They have the capacity to fulfill the roles and live up to their responsibility and objectives taking into account the conditions of the Strategic Initiatives.

The Strategy Governance Team with the Change Control Board and the Process Governance Teams all know how to initiate and perform PQA to improve success factors and change direction of the Strategic Initiatives if and when this is required.

However, how does this organization become aware of the need for change?

7.5.1 The Coffee Bean Methods

For strategy implementation, we will look at the method elements from a process point of view with development, implementation, and project and quality management that bring us all the way from the original idea and need for change to the final use and operation of the developed and implemented solution (Figure 7.3).

The idea behind the coffee bean strategy of simultaneous Development and Implementation coordinated by Quality and Project Management is to promote systems and lateral thinking and synergy leading to results that exceed the expectations of the stakeholders. It is an agile strategy.

The method that we have been using comprises the following elements:

- All Strategic Initiatives produce their results based on a requirements specification that is sufficiently detailed to guide the Workgroups in their development, implementation, project management, and quality management work.

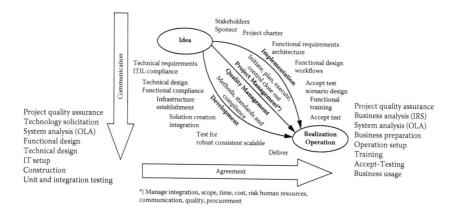

FIGURE 7.3
The Coffee Bean Model.

- All Strategic Initiative-work close out with Accept-Testing that proves that the latest version of the requirements specification, which is the one comprised in the latest baseline, has been adhered to.
- Communication:
  - Progress measured by KPI.
  - Deviations from baseline measured by KPI, but also measured by result quality, changed conditions, and pertinent events.
  - Strategy and Strategic Initiative evaluation with Process Governance Team and with Strategy Governance Team ensuring ongoing sponsor support.
- Development of functional (error free) fully normalized databases and processes using agreed technology and ensuring the availability of a feasible information system operation and support.
- Implementation of fully functional solution elements ensuring availability of business operation support facilities such as user guides and training material.
- Project and quality management that ensure availability and efficiency of resources for development, implementation, and ongoing simulated Accept-Testing of solution components delivered from development and implementation. Project and quality management survey and evaluate the Strategic Initiative progress primarily based on simulated Accept-Test results, but also looking after organization and process issues.

7.6 STRATEGY GOVERNANCE

We have planned and initiated our Strategic Initiatives with a solid requirements specification and documented result expectations agreed to by sponsors and other key-stakeholders, but how can we make sure that we actually get what we want?

Our plans and our requirements specifications are predictions; it is pure luck if we get a solution that fits the requirements specifications unless we manage the solution delivery.

The Strategic Initiative does not only plan and execute projects; it also changes business behavior, re-organizes organizations, and invests in the development or acquisition of new products and improved methods where the outcome depends on how the market and the internal stakeholders accept the change.

This makes the outcome of Strategic Initiatives random and for both forecasting and tracking we need to apply statistical methods in order to understand the magnitude of variance with which we are faced.

In the Strategic Initiative world of random outcomes, agile planning and tracking of our activity allow an easier to understand follow-up on progress. By measuring progress only by user- and management-accepted solution components delivered, installed, and ready for production we know for sure what we have and we have a steadily improving foundation for estimation based on the gained experience.

The benefits from agile solution delivery in terms of solution quality, process efficiency, and cost reduction far outweigh the planning effort and the scenario establishment investment for their realization.

7.6.1 Negotiation

Strategic Initiative governance is communication and negotiation. The Governance Teams and the Workgroup Teams get and deliver information that indicates that planned activity is not executing the way it was planned:

- Resource availability is not as promised.
- Resource skills and experience are not what was contracted so important activity does not yield the result quality expected and it takes too long to get the results.
- Critical activities cannot start because they wait for resources to be released from other activities.
- Critical activities are interrupted because important resources leave or because development components from the COTS vendor do not work as expected.

I have never been involved with Strategic Initiative projects and programs that do not have these problems.

Risk management can just tell you the threats might happen and that you need to plan for this eventuality with appropriate risk response.

To avoid these threat events is only possible if you double or triple the involved resources and with them the cost of the project and the required budget.

By allowing a little more time for delivery, you might have time to adapt to the problem without major cost increases and still have happy stakeholders.

These alternative possibilities for avoiding or mitigating the risk demand that the Strategic Initiative sponsor and Governance Teams negotiate with solution delivery Workgroups of internal and external resources in order to find the best possible way to handle the risk situation.

7.6.2 Estimation and Forecasting

PQA is our first foundation for estimation and forecasting when we plan a Strategic Initiative and when we adopt major changes to the Strategic Initiative, that is, when we re-plan the initiative:

- The success factors give us targets of tangible and intangible nature that can be used for negotiation about where to go and for asking questions as to whether a success factor has been achieved.
- The success factors give us an idea about what to implement and why.
- By showing the expected tangible and intangible values of needed solutions and the planned activities to perform their delivery, the success factors provide a base of reference for prioritizing and evaluating the requirements specifications for the solution components to be delivered by the Workgroup Teams.
- The PQA activities with their outline estimated duration and resource usage give us the foundation for more exact estimation and establishment of milestones and baselines against which we can measure progress.
- The milestones tell us the expected delivery date of solution components so that we can have requirements, test scenario, and people ready for simulated and final Accept-Testing.
- Other baseline elements such as the critical path of Workgroup tasks can tell us about the probability of delivery of the final result on planned time.

The forecasting of the product quality is the requirements specification. We measure the quality of delivered solution components against the requirements specification that has been broken down into very specific and tangible test cases and expected test results.

The breakdown of the requirements specification in use cases and workflows prepares for an agile delivery process with agile planning and tracking.

7.6.3 Strategic Initiative Key Performance Indicators

Once initiated, the progress of the Strategic Initiatives is tracked in order to be able to adapt to new knowledge and other changed conditions.

Over time the risk profiles change and with those the probability of opportunities and threats.

The tracking is based on KPI that can provide information about:

- Organizational condition changes
- Solution condition changes
- Process condition changes

Some KPI are specific for each condition type, while others look at cross-initiative environment indicators that indicate if duration or cost is under control for the Strategic Initiative.

You will also need KPI that can tell you whether your Strategic Initiatives perform as they should from a business perspective, that is, looking at all opportunities and threats known at a given point in time:

- Is what we are doing still attractive?
- Do we need to change the strategy and re-establish more valuable Strategic Initiatives?

A KPI that can help with this tracking is the Compound Expected Value (CEV). Calculation of CEV per Strategic Initiative makes it possible to compare the initiatives mutually and to evaluate them in the context of overall strategy performance.

7.6.4 Communication

We need to communicate in order to keep all stakeholders happy all the time or at least to make sure that the stakeholders understand why they have a reason to be unhappy.

We have used several types of communication:

- PQA
- Meetings for initial scope definition
- Presentations for PQA and IRS participants
- IRS interviews

- OLA-based IRS consolidation
- 9:00 meetings for progress check and fast reaction to issues
- Change request tracking from Workgroup to Strategy Governance Team
- Status meetings to announce major changes in direction and objectives

All of this communication is planned for and supported by standardized documentation and presentation formats that can be adapted to specific situations or improved over time as experience is obtained.

The objective of the Workgroup Manager communication is to ensure that "no excuse for failure" is obtained through communication and negotiation:

- With the Governance Team, you want to obtain whatever scope changes are needed with respect to funding, time, and solution quality. This negotiation is based on what you know from dialogues with team members, the issue and error documentation, and the Workgroup KPI.
- Within the Workgroup, you ensure that development and implementation teams communicate in an agile way.
- You listen to Workgroup resources in order to ensure efficient solution delivery.
- You call the Workgroup Team together to inform them about needed changes.
- On a daily basis, you meet the active Workgroup Team Members in person-to-person dialogues and in 9:00 meetings simply to keep the team spirit high, but also to discuss issues and problems of both personal and work natures.
- With internal and external stakeholders, you negotiate the resource needs and ensure that these resources are informed and motivated to work in or be used in your Workgroup Team.

IRS and operation of the Planning Information System can be left with other resources such as facilitators and coaches from the Project Office and a Program Office:

- Once the project plan has been estimated, the Project Office can make sure it is registered in the Planning Information System for correct tracking and reporting.
- The Project Office can ensure and control that Workgroup Team Members add task status information in the form of person-hours

worked on tasks and work effort estimates in the form of person-hours to complete the task. As Workgroup Manager you are still responsible for reliable time and cost to completion forecasts because this is a result of negotiation.

- Project Office can ensure that the Workgroup Manager gets the pertinent information for plan adaptation and KPI generation as required by this manager.
- IRS and OLA are best performed by professional business analysts and facilitators in the Program Office, not by Workgroup Managers.

Management is about communication and negotiation in the context of planning and tracking of the Strategic Initiatives.

Appendix A: PQA Introduction to Participants Example

Place L-9999 XXXXXX
01-02 June, 2013
Start 900
Finish 1800

PARTICIPANTS

Peter, XX IT manager
Paul, XX Trade desk manager
Mary, XX System support manager
Carl, XX IT Infrastructure manager
John, XX and YY Security manager
André, XX Project manager
Andrew, YY Project manager
Christina, XX and YY Network manager

FACILITATOR

Soren

A1. INTRODUCTION TO PQA

Process Quality Assurance (PQA) is a team-building technique for groups of people who are going to cooperate in order to solve a complex task, for example, a Strategic Initiative.

The scope of the task and the group of people to participate in the PQA process is decided by the task sponsor, who must be supported be the PQA facilitator. The result of this decision is a PQA Introduction such as this one, showing core quality objects.

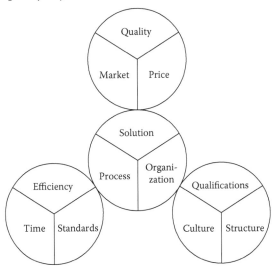

PQA visualizes the complete set of agreed requirements to the solution, the process, and the organization resulting from solving the complex task:

- The initial scope and conditions of the task signed off by the sponsor
- The visions and missions of the involved key-stakeholders
- The Success Factors
- The Critical Success Factors (CSFs)
- The required activities to achieve the Success Factors
- The organization to perform the activities
- The solution structure and framework

The PQA process is initiated in a PQA workshop. The PQA workshop makes visible the full scope of the task:

- The initial scope and conditions of the task signed off by the sponsor
- The key-stakeholders participating in the workshop

- The visions and missions of the involved key-stakeholders
- The Success Factors
- The CSFs
- The required activities to achieve the Success Factors
- The workshop participant to facilitate an activity

PQA is one tool for project management, but it does not replace good project management practice.

A1.1 PQA and Destructive Conflict Prevention

If conflicts could be avoided among project participants and between project sponsors and project performers and other involved stakeholders, such a project would have a very high chance of producing a successful outcome.

In order to avoid destructive conflicts, all project participants must be highly motivated for both personal reasons and for creating mutual success.

Such a situation can be achieved by promoting common targets with a related benefit package, which can only be obtained if all participants reach the target simultaneously. Such a situation will motivate and even force the partners to help each other instead of fighting.

Why are all projects not handled in this way?

Because it requires quite some strategic thinking and planning to establish such targets.

When building a bridge common targets are quite obvious, which means that the "no conflict" solution is more process and organization based, which in most cases can be controlled by management.

When building IT infrastructure and information systems in support of private and public business activity, the common targets are much less obvious.

Quite often the resulting solution from such projects does not reveal itself until the final acceptance testing. It therefore becomes a key to the "no conflict" solution that targets are thoroughly defined and agreed to by all project participants and other stakeholders. This holds true especially when we talk about "moving targets," which is more the rule of the game than an exception.

The project process must be thoroughly defined and the project organization put together in such a way that the "no conflict" solution can and will be reached and communicated. There must be clear rules for how known and unknown risk events and problems are treated organization- and procedure-wise.

PQA ensures a strong team feeling based on mutual respect and commonly agreed decisions within the project organization, which also strengthens its ability to cooperate with external partners. In this way PQA contributes to establishing a situation with less risk of destructive conflict. Nevertheless, the change management procedures while conducting the complex task must be clearly defined as well as involving external stakeholders, which is a strong requirement for the defined activities under PQA:

- The situation without destructive conflicts is a success factor.
- The change management process is an activity on its own or imbedded in other activities such as project management and quality management.

A1.2 The PQA Product

The PQA Workshop results in a documented outline project plan with agreed result requirements and outline responsibilities. The complete project plan is not detailed defined, estimated, and scheduled until immediately after the PQA Workshop. Depending on the organizational level of the PQA participants (PQA can be used by top management for 5-year planning or by a task force for a single project), the outline project plan can result in several projects requiring their own PQA Workshop for team building and detailed planning, or it can simply result in an agreed task list.

PQA standard documentation ensures a result that is easily interpreted and understood. It is a good foundation for later status reporting, project reviews, and project evaluation. The PQA Workshop documentation is used for:

- Introduction of the project to future project participants.
- Introduction to management, who will deliver required resources to the project.

The PQA documentation produced after the workshop visualizes estimates, budgets, and risk.

A1.3 The PQA Organization

The key-stakeholders to participate in the PQA Team are selected based on criteria such as:

- Competence within the scope
- Relevance within the scope
- Completeness of knowledge pertinent to the scope

Together the PQA Team participants cover the complete knowledge and experience required in order to solve the task at hand.

Even though the participants in the PQA Workshop might not possess all required skills and competences themselves, they must be able to evaluate the skills and competences required to complete the task.

The participants in PQA are at best selected from comparably equal levels of management, responsibility, and authority within their respective organizations. During PQA, all team members are peers irrespective of their basic organizational level in order to be able to express their ideas unbiased by management pressure.

The PQA participants are trusted by the project sponsors in order to be able to get approval of their project plan.

A1.4 The PQA Process

PQA requires maximum active participation from each participant. It is impossible to be a passive passenger in the PQA process. By "forcing" the participants to listen actively and respectfully to each other, it is ensured that every participant achieves maximum inspiration from the other participants.

All participants get to understand and respect their own and the others priorities, requirements, and wishes. In this way, the project group builds a mutual understanding of the project scope, complexity, and target, which can be communicated to interested parties outside the PQA participating core group.

PQA ensures synergy.

The PQA process ensures that resulting documentation is produced "on the spot," which reduces the risk for later manipulation by a creative author of minutes. The result reflects precisely the agreed conclusion.

Documentation, which requires specific skills and competencies for its production, is done after the Workshop by involving people with the necessary skills and competencies. Such documentation encompasses activity descriptions for each activity outlined during the workshop.

A2. THE SCOPE OF THIS PQA

In order to be able to control the complete XX IT environment of hardware (mainframes, PCs, routers, networks), software (operating systems, databases, TP monitors, middleware), and applications (end users and Back Office), this environment must be identified, managed, and controlled.

Identification means that any relevant item can be overviewed from a single point of reference based on access to central or distributed information about the whereabouts and condition of the item.

Management and control means that no item can be in conflict with the overall objectives of XX concerning availability, reliability, safety, and maintainability of applications and solutions. Whenever an item is one of the keys to the performance of a user application, it must be ensured that any deviation from expected item behavior is discovered and dealt with.

During this PQA, the participants shall define the solution, process, and organization requirements concerning the establishment of the XX IT enterprise management within the framework of the YY project already executing.

During the PQA, you will get the opportunity to describe and get documented how you think the XX IT infrastructure success can be ensured. You will get inspiration from and give inspiration to the other PQA team members.

As a result, we will have an outline description of all the necessary and sufficient activities, the organizational preconditions, and the complete solution.

A2.1 The YY Project Situation

The YY project is delivering a fully functioning web-based bank supporting private financial transactions performed by the bank's clients, and transactions performed by the bank with respect to stock exchanges and related services from other systems such as SWIFT and cooperating banks.

Although some functional experience can be drawn upon from the current private banking system, the YY system must be considered a completely new application based on new technology.

The YY system is currently in Acceptance Testing state, which should imply that most solution components have been developed, tested, and approved on their own and integrated with other relevant components. This is not quite the case, which is a major risk and challenge concerning the future performance of the YY project.

Until now, the requirements concerning XX IT infrastructure support have only been outlined on a very high level. The current YY test environment in XX is by no means representative of the future YY IT production environment. All testing is currently concerned with business functionality, not with IT environment performance or availability.

A2.1.1 Many Developers' Results To Be Integrated and Approved in Parallel

The YY solution is designed and developed by many individual contributors and vendors. Each vendor's components are going to integrate and interact with other vendors' components.

Until now it has been impossible for YY to obtain documentation for how the different components are going to interface with each other and what processes and applications are going to be interfaced, used, developed, handled, and controlled by XX.

The YY Acceptance Testing (SAT) is taking place in parallel with Proof of Concept (POC) testing ad hoc using IT infrastructure elements at XX. The testing results in ongoing delivery of new or enhanced or corrected components from the different contributors to the YY solution. This situation makes it very difficult to guess the final outcome of the processes and components that must be delivered by XX.

A2.1.2 XX Not Actively Involved in Project Planning

Because the focus of the YY project until now has been business functionality and because the SAT testing situation is less than optimal, there is very little time devoted to planning of the XX IT infrastructure implementation.

It is of course important that the business functions are consistent, valid, and user friendly before they are launched.

But it is to a very high degree the IT infrastructure that shall ensure whatever availability, reliability, and safety that will be required for the final YY solution.

It is therefore a very high risk to the whole YY project that the XX IT infrastructure is not properly defined, documented, and developed at this late stage of YY SAT testing, especially regarding the lack of experience within XX with respect to the established or expected YY technology, transaction quantities, and transaction frequencies.

Before launch it must be proved that the YY solution is technically robust and well performing.

A2.2 The YY Enterprise Management Opportunities

Available Enterprise IT Management System software can allow XX IT to establish the IT infrastructure that will be required by the YY solution.

Standardized Agent and Manager Software from equipment manufacturers and from CA can allow XX IT to see the most relevant events and to react to these correctly and in time in order to keep the business applications available with the highest possible performance.

XX IT or its vendors can build customized agents and managers for direct surveillance and control with all critical processes not otherwise managed.

Time critical actions in response to events can be automated.

Workload Management (job control), Storage Management (back-up), Help Desk, and even contributions to end user solutions form natural parts of the future overall XX IT Enterprise IT Management System Strategy.

The establishment of the future Enterprise IT Systems Management information system solution is almost as complex as the establishment of the YY business functionality and customer service. Please refer to the attached "Enterprise Systems Management" whitepaper from XX IT.

Only a close cooperation between business interests and IT interests within the YY project can ensure a future complete XX IT Enterprise IT Management System solution that performs according to YY and XX IT expectations.

A2.3 What Are We Going to Achieve in the Workshop?

In the PQA workshop we shall consider all the outstanding activities to be performed by the YY project team in order to ensure delivery of the full YY solution for commercial usage.

We shall formulate why and how and when close cooperation between YY business-processes implementation and XX IT infrastructure implementation can bring about the optimal YY solution.

The participants in the workshop must formulate the requirements concerning:

- The future working condition in the YY organization with complete XX IT Enterprise IT Management System solution implemented.
- The adjustments to the YY project organization, processes, and products, which can ensure an efficient cooperation between all future YY project participants, as well as considering the involvement of external vendors.

We will not formulate specific technical requirements to the future solution as these requirements will be precisely defined by competent resources task by task later on.

However, we will define and visualize the targets for these competent resources to be able to work effectively and produce efficient solutions that can and will be accepted by XX IT and YY management. Relevant ideas and suggestions are welcome.

A2.4 How to Prepare for the PQA

Please read the "Enterprise Systems Management" whitepaper from XX IT before the PQA Workshop.

Based on ideas formulated in this paper and on your own ideas and expectations concerning the implementation of the future PFS production environment of systems facilities, support, standards, and environment administration, you must describe your vision of the future YY and how you interpret the mission of the YY project team.

Think about what you consider the most extreme case of success for the development and implementation of YY and the XX IT infrastructure. Describe the factors especially relevant to your responsibilities and capabilities within the YY project team.

A2.5 Areas of Concern

The following areas of concern should be considered:

- Are deliverable success criteria defined?
- How are reliability, maintainability, and availability defined?
- With what other systems do we integrate?
- Who is the user of the product?
- How should the XX IT infrastructure communicate with the users?
- User involvement?
- The implementation process?
- Use of methods, techniques, and tools?
- Availability or development of standard documentation templates?
- How do we ensure compliance with standards?
- What standards do we have to develop or implement?
- Education, training, and coaching requirements?
- Common reasons for IT failures and their prevention requirements?
- What is the biggest challenge concerning Back Office operation?
- What are the most common recovery problems encountered and what are their prevention requirements?
- Who should participate in the implementation of the XX IT infrastructure?

See you in XXXXXXXXXXX.
Yours Sincerely,
Peter Soren

A3. PQA AGENDA

A3.1 Verbal Introduction to the PQA

The verbal introduction to the PQA comprises the same information as this written one, but allows for a discussion of the scope of the PQA Workshop. Also it permits the PQA facilitator and the PQA sponsor to present who they are and to explain their own and the participants' roles during the Workshop.

A3.2 Definition of Vision/Mission

Definition of vision/mission is done individually by each PQA Workshop participant. The participant presents a personal view on the expected result and the expected business benefits.

Each individual vision is written down with the name of the author attached to it. It is one of the keys to the cooperation and mutual respect of the team members that they understand the motivating factors of the others.

It is allowed and even recommended to ask for explanations of vision/mission statements, but their relevance or correctness can never be challenged.

People will and shall tackle problems in different ways. We just have to know what the cases within the PQA Team are in order to be able to play on each participant's strengths in order to obtain maximum creativity and synergy.

A3.3 Suggestions for Critical Success Factors

The Success Factors express what the PQA Team thinks should be the quality of the result of its work. The Success Factors also express by what the group expects that other people will evaluate the project result.

The Success Factors are identified by letting each participant in turn suggest one Success Factor. All suggestions must be true, relevant, and valid, but full unanimous agreement is not required.

This suggestion process is continued until no one has more suggestions. Normally 40 to 60 Success Factors are identified.

A3.4 Definition of the Critical Success Factors

The suggested Success Factors are grouped into CSF classes in such a way that the suggested Success Factors belonging to a class define the class's CSF expression. There will be from 5 to 9 CSFs. Full unanimous agreement of all PQA Workshop participants is required on each CSF formulation. All suggested Success Factors must belong to at least one CSF class.

A3.5 Outline of Activities

This process produces the CSF matrix shown in Table A.3 under Section A5.2 that shows how the outlined activities contribute to the fulfillment of the CSFs and their Success Factors. Participants in turn can suggest any number of activities which they believe are required to ensure the realization of the CSFs and their Success Factors. Each activity supports at least one CSF. How the activity will contribute to the fulfillment of the CSF and its Success Factors is defined after the Workshop.

The facilitator negotiates the level on which the activities are defined in order to ensure that a too detailed activity level or too high activity level is not reached. The defined activities must be agreed to be on an equal level in the future Work Breakdown Structure (WBS) that will be defined after the Workshop.

When no more activities are suggested, each suggested activity is cross-checked against the CSFs, which it supports. In order to control the completeness of activities, each CSF is then checked in order to verify that the suggested activities together are adequate to ensure fulfillment of the CSF and its Success Factors. Missing activities are added and evaluated against the CSF.

During the activity definition session it is important to ask the suggesting participants to define exactly what they mean by a suggested activity unless it is absolutely self-explanatory; however, this is rarely the case. After the workshop, each activity must be exactly defined by the person who becomes responsible for the activity. It is highly recommended to take notes during the Workshop in case you become the responsible facilitator for an activity.

A3.6 Assignment of Responsibility

Assignment of responsibility for the definition, risk evaluation, estimation, and planning of each activity is next. Assignment must be given to a participating PQA Team member, *never* to an external organization unit. After the Workshop, each assigned team member must within a set time frame produce a complete activity description with activity definition, task list, resource requirements, time and cost estimate, risk evaluation, and plan information about predecessor and successor activities.

A3.7 Evaluation of the Quality by which the Activity Might Already Be Carried Out

Some suggested activities might already be executing or defined by another PQA Team, while others are new to the PQA Team. Activities that are performed satisfactory are of less interest to the PQA Workshop, but the PQA Team needs to be aware of this. Other activities should be evaluated according to their current performance in the organization. Their "value" is stated as:

    0   Activity not done
    1–2   Activity known but unstructured performance
    3–4   Activity performance structured, but to be improved
    5   Activity probably satisfactory performed

A3.8 Review Planning

It is absolutely necessary to decide on the date of a review meeting with all PQA Workshop participants attending. In the review meeting, all Activity Descriptions are approved before the final WBS and schedule are drawn up.

A person must be appointed to be responsible for coaching the authors in their usage of the Activity Description form and for collecting and distributing all Activity Descriptions at least one week before the review meeting.

This person will also organize the rewriting of the PQA Workshop result on standard forms.

A4. AFTER THE WORKSHOP

A4.1 Detailed Define and Estimate Activities

Each person assigned as responsible for activities makes an activity description for each of these activities. When doing this it is important to assess all thinkable risks concerning estimates. Availability of resources with the proper skill and experience or the ability of sub-contractors to deliver the required quality on time are the most common risks in software development projects. If new technology is involved, the ability of the organization to get and control it is a major risk area. The consequences of risks should be explicitly stated in the description of involved activities.

A4.2 Build the Work Breakdown Structure

The activities are sorted into their natural sequence and they are grouped according to the WBS groups of work, for example:

- Implementation activity
- Development activity
- Quality Management and Project Management activity

These activities are further broken down into specific deliverable elements.

In other cases, WBS can be organized by deliverable component, where each deliverable component has activities for implementation, development, and quality management.

In this process, milestones are identified and agreed to by the PQA Team and the key-stakeholders and sponsors.

A4.3 Build the Project Plan

After having organized the activities into a WBS with phases, activities, tasks, and milestones, the next step is to define dependencies between tasks and allocate resources to tasks.

The last step is to optimize the plan in order to meet milestones and deadlines within resource constraints and cost budget.

The project plan and the complete set of PQA documentation is approved by the sponsors and the key-stakeholders before it is distributed to other participants.

A4.4 Initiate the Work

The work can be initiated when all project participants have received the approved plan and PQA documentation.

Before initiation it should be decided what kind of information is needed by the project and what kind of information is needed from the project. The decisions should be documented in a distribution list for the documentation and in a documentation description showing the structure and content of the information as well as who is responsible for its production and its distribution.

It must be assured that the project is properly managed within the project management rules of the organization. Quite often these rules have to be established and documented explicitly within each project. The general rule is that every project gets its own culture, which implies a need for its own procedural handbook.

A4.5 The PQA Result

The PQA results produced directly in the Workshop is a set of documentation comprising:

- The individual vision/mission statement of each participant
- The CSF and their definition (the underlying suggested Success Factors)
- A cross-reference matrix between CSF and activities, which also shows activity evaluation and assigned responsibility

This documentation is transformed into standard PQA documentation for distribution to the participants immediately after the Workshop.

After the Workshop the PQA activity list is defined in more detail using:

- Activity descriptions
- A WBS
- A schedule
- Document definitions
- Document distribution list

Appointed project managers are responsible for the detailed project documentation creation, the Project Requirements Document (PRD). The PRD makes what is going on visible to key-stakeholders, sponsors, resource management, employees, involved resources, and other stakeholders.

Vision Examples

Visions	
Eric Pavier:	We want to be outstanding with respect to usage and implementation of methods and techniques.
John Doe:	Our consultants are rated the best when it comes to project management and analytical skills.

The PQA process has ensured that the involved participants and interested stakeholders have reached a common understanding of the future project.

The full set of PQA documentation is approved in a review. It will be used for later evaluation of the project progress and its results.

A5. PRODUCED IN THE PQA WORKSHOP

A5.1 The Vision Statements and the Critical Success Factors

The vision statements are listed with the name of the author.

The CSFs are anonymous because most of them are suggested based on the inspiration from all of the team member suggestions. The CSFs are documented like this:

CSF and Success Factor Suggestions

	CSFs
1.	A supply service that can be measured and optimized
	On-time delivery to customers
	Avoid or reduce the back-order shipments
	We understand what good delivery service is (it is defined)
	Improved availability of goods due to better forecasting
	Efficient support for the customer and the salesperson in the order process
2.	A competent organization
	Simultaneous adaptation of business, management, organization, and system procedures
	We need to build an organizational match before installation of the system
	The users involved are the best we have
	Ability to handle big, complex, international projects

A5.2 The PQA Matrix

PQA Matrix

Critical Success Factors						Value in execution	Responsible for planning
5. Quality management							
4. The system supports a dynamic business environment							
3. Accessible information							
2. A competent organization							
1. A supply service which can be measured and optimized							
Activities	1	2	3	4	5		
Define all aspects of a good supply service	*	*			*	4	LH
Build a development support organization		*			*	2	JD
Build the user competence necessary to utilize the new system		*	*	*	*	3	LH
Define the user competence necessary to define the new system	*	*	*	*	*	2	JD
Distribute the IRS report		*		*	*	1	PP
Inform involved sales companies about the process and the progress ongoing		*			*	1	PP
Define and build the complete system for communication, HW, SW, and applications	*	*		*	*	0	CV
Do PQA with the users involved in the design	*				*	0	CV

A6. DOCUMENTATION PRODUCED AFTER THE WORKSHOP

A6.1 The Activity Description

The Activity Description is done for each activity defined in the PQA matrix.

Activity Description	Activity ID:
By:	Date:
Scope	Describe why this activity is required and what its areas of concern and responsibilities are.
Deliverables	A description of the expected outcome, e.g., a tender material, a report, or an accepted system.

Purpose	A description of the deliverable quality expectations or of the expected benefits from the delivery of the deliverable.	
Responsible	The person responsible for getting the activity done (sometimes the person writing this description).	
Resources/Roles	A description of needed roles and their responsibilities and skill and competence requirements. Specific named resources can be applied to the roles.	
Task List	**Task Description**	**Estimated Resource Usage**
	Scope purpose and product for each task. Required role interaction if not obvious.	Roles/names of resources to perform task with person-hour estimate for each resource.
Time frame	Duration or fixed period.	
Risk Assessment	Describe potential events or preconditions that could influence the deliverable quality, the duration, or the estimated resource usage. Think about external factors, which cannot be controlled, and internal factors, which can and must be controlled. Describe the probability of the event (if 100%, it is a problem rather than a risk). Describe the impact of the event.	
Dependencies	Reference to activities that must be performed before this activity is performed and to activities that will use results from this activity. Explain if not obvious.	

A6.2 The Risk Response Matrix

The Risk Response Matrix is used for control of the completeness of the risk response strategy established while the PQA participants are reviewing the Activity Descriptions produced after the PQA Workshop. The project manager or the PQA facilitator produces the risk list shown in the second row (Risk). The third row shows for each risk the originating activities. In the example, the WBS reference is shown.

During the review, the risk responses are agreed to and listed in the first column. The second column shows who is responsible for executing the risk response, the third column shows the deadline for the response execution, and the fourth column shows the activities involved with the response.

Risk Response Matrix

Project Name				
(Who, Date, Version, Distribution list, Approved by ....)			**Risk**	**1. Availability of skilled resource**
Response activity	Responsible	Deadline	WBS-ID	2.1.2, 3.1.1
Procure sub-contractors			3.1.1; 4.1.1	*
Hire Java Ace			2.3	*

One response can have a positive or negative effect on more than one risk event, which is shown with a * in the cross-reference cell. One risk event might require more than one response.

The PQA participants ensure the best possible response strategy by controlling that set of responses to a risk is the best possible way to avoid, mitigate, transfer, or share that risk, and by controlling that the full set of impact on all risks by a given response is fully understood and documented.

A6.3 The Work Breakdown Structure

The WBS shows the hierarchy of the project activities.

Here it is shown as indented activities in a Gantt Chart, which also shows the project plan schedule (Figure A.1).

A6.4 The Work Breakdown Structure Dictionary

The WBS Dictionary contains all initial scope information concerning the project, the program, or the strategy execution planned, which comprise:

- The initial activity scope such as formulated in the PQA Workshop Introduction.
- The full PQA Workshop result with participants vision/mission statement, CSFs related to their detailed success factors.
- The PQA matrix with initial responsibilities and activity quality shown.
- The risk response matrix.
- All activity descriptions.
- The WBS shown graphically as in a Gantt Diagram:

ID	Name	Start	Finish
EM	Enterprise Management	15-05-00	29-06-00
EMP	Planning	15-05-00	29-06-00
EMP01	PQA Preparation (Org/Case)	15-05-00	15-05-00
EMP02	PQA Workshop (Agreement)	17-05-00	18-05-00
EMP03	Project Planning and Management	19-05-00	28-06-00
EMP04	Status Reporting	22-05-00	28-06-00
EMP05	Change Reporting	24-05-00	28-06-00
EMP06	Management Reviews	24-05-00	29-06-00
EMB	Business Case Development	15-05-00	12-06-00
EMB01	Cost Benefit Analysis	15-05-00	29-05-00
EMB02	User Documentation (Workflow/Training)	15-05-00	12-06-00
EMR	Results Production	25-05-00	15-06-00
EMR01	Document results	25-05-00	05-06-00
EMR02	System Documentation (LOG, Explanation)	03-06-00	15-06-00
EMT	Testing and Delivering	02-06-00	29-06-00
EMT00	Establish Test Scenario	02-06-00	15-06-00
EMT01	Simulated Acceptance Testing	02-06-00	24-06-00
EMT02	Final Acceptance Testing (Sign Off)	23-06-00	29-06-00

FIGURE A.1
Gantt chart schedule.

The WBS Dictionary is the primary project description, and it is an important base for the project plan.

A6.5 The Communication Plan

The communication plan shows the type of information that will be produced in the lifetime of the project, the program, or the strategy execution.
For each information type it is shown:

- Who is responsible for producing the information. Who is responsible for approving the information before it is made available to the receivers.
- The document standards to follow or the form of the information suggested, for example, minutes of meeting, presentation on Web or in meeting, training material, solution documentation, etc.
- The technical whereabouts concerning the information, that is, where it is stored, in what format, how it is protected, how access is ensured, etc.
- What is the content of the information and what are the origins and the required quality of this content? The quality is what the receiver expects to see and how this quality is ensured.

- The frequency of the information or the events that trigger the production and delivery of the information.
- The receivers of the information or, alternatively, who must have authority to view the information.
- How, who, and when to inform the receivers about the availability of the information.

The communication plan is at best prepared as part of the Activity Description review process.

Appendix B: PQA Workshop Result Example

PQA Workshop Result
20. – 21. May 09.00-19.00/14.30

PARTICIPANTS

Development Manager
Financial Manager
General Manager
Client Project Manager
Sales Manager
Production Deputy Manager
LI Project Manager
Production Manager
Quality Manager
WCAT IT Project Manager
Client Service Manager

FACILITATOR

Soren Lyngso

B1. INITIAL SITUATION AND OBJECTIVES

The company is one of the world's most important suppliers of systems for production and maintenance of catalogs to be reachable from many media. The catalogs can retrieve information across many databases and data media from alternative data providers.

Customers are typically large international companies with internationally distributed production and sale of their products.

Projects comprise internal development of standard systems and tailor-made customer solutions. Client solutions may be recurring orders involving only a few adjustments in each case, or they may be completely new orders with varying degrees of solution development.

Customer projects today require no great intensity in cooperation with the customer, but there is a tendency that it is becoming increasingly necessary to involve the client's employees in system development, especially when it comes to more advanced user interfaces.

The wish for a new and better project environment must be seen as a natural progression toward greater efficiency and steadily improved competitiveness and customer service.

To improve efficiency, a better overview of production processes and their contexts is needed. Production process data must be collected systematically in a way that allows for preparation of standard times for standardized procedures. Production process data must show how cooperation between the various departments is handled during a project so that handover of partial deliveries can be ensured in time with the right quality.

It must be possible to assign available scarce resources across national borders.

It must be possible to prioritize projects, so that less critical projects do not inadvertently get critical resources assigned at a time when a more critical project requires their usage.

In order to increase competitiveness, besides through increased efficiency, it must be possible to analyze the impact on the production capacity of each customer order accurately in order to be able to build a realistic plan for cooperation with the customer on completion of a delivery.

The teams that are expected to complete a project must be shown as early as possible in the project database so that departments can plan their resource availability. The overview of resource capacity and the status of ongoing projects must be internationally available to sales and project management.

The future project information system must create visibility of active and planned projects, so that the cooperation between project participants, project managers and department managers can be handled on a realistic basis.

Acquired experience must be classified and available so that it can be used for future estimation and projects planning.

It is especially desirable that project participants can visualize that they have been good at planning their projects.

The project management information system combined with improved business processes must ensure reliable and consistent information and communication in support of the daily project management and the long-term capacity planning of resources.

It is essential that the system is implemented in harmony with that the organization is ready to exploit the system. Many small victories are better than one "big bang."

It is important that workflows around the system usage are in place and that there is education covering the concurrent usage of all integrated information systems for project management.

Finally, it is important that pertinent standards are established and implemented before the systems go into operation.

B2. VISIONS AND MISSIONS

B2.1 Development Manager

Based on our "Single Source" project database, employees at all levels have access to information that is sufficient to enable them to prioritize correctly. We get a collection of documented experiences that efficiently supports continuous improvement of our methods.

B2.2 Financial Manager

We get one common methodology across national borders.
This gives overview and visibility of impact on the project, department, and company level.
Efficiency and higher profits.
Not too bureaucratic.

B2.3 General Manager

A system that provides a reliable overview of responsibilities, milestones, deadlines, and status in the planning and execution of projects and that creates transparency and mutual understanding of processes and organizational relationships.

B2.4 Client Project Manager

We get a visible international project management system that is easy to maintain. The system facilitates an easier daily follow-up for management and project managers.

Project plans are maintained in a central location with real-time data that is accessible to stakeholders.

- Internal and external project plans are maintained in one place.
- It is visible, what activities are new compared to what was agreed.

B2.5 Sales Manager

Customers benefit directly from new solution:

- Web access to project status
- Visible development in cost and invoice base

Competitive advantages in the sales phase:

- Fast calculation and time estimation
- More solution options

Improved operational execution:

- Higher efficiency
- Improved tracking and experience documentation
- Common methods

B2.6 Production Deputy Manager

The system is reliable and shows the true picture of the project portfolio and the capacity utilization.

The system supports both short-term impact assessment and long-term capacity planning.

Potential projects get visible in the plan.

B2.7 LII

All stakeholders are satisfied with the solution.
The solution is used as intended.

B2.8 Production Manager

Visible project execution (processes and resources)
Visible dependencies
Overview of project progress
Only three people update the project plan

B2.9 Quality Manager

We get an overview of executing and agreed projects and tasks so that each employee can make decisions about where the employee can set in with maximum benefit. This gives better job satisfaction, creates greater flexibility, leads to fewer errors, and results in higher throughput and improved performance.

B2.10 IT Project Manager

We establish a solution that is used.

B2.11 Client Service

The system supports both manager and employee to better planning and monitoring of projects based on "Single Source" available project information.

B3. CRITICAL AND SUGGESTED SUCCESS FACTORS

B3.1 Effective project culture

1 Impact → only projects in the system receive resources!
3 The objectives of the system are known and accepted.
5 Planning culture with respect for others planning.
6 Provides insight for others into my projects.
13 Responsibilities and tasks are always clearly defined.
14 The users understand the benefits and importance of the solution.
15 The system is embedded in the organization—also in 2 and 3 and *n* years.
18 Increased employee satisfaction through better coordinated production processes.
21 The system must support the sharing of good ideas.
27 We get a helpdesk for users of the system.

30 Full overview of resources and their qualifications and their availability across the entire international organization.

32 More efficient project process.

40 It becomes apparent who must be notified when there are deviations from a plan.

42 All involved stakeholders report problem situations in the system with the assurance that the situations are treated in time.

43 Most activity in WCAT happens in projects and all departments are working toward common goals of the projects.

44 The methodology foundation becomes so good that we can live a week without the system in operation.

46 Who is responsible to update what in the system is visible.

47 No excuse for not being proactive and for not taking initiatives.

49 An internal team at WCAT is identified to ensure the success of the new system and the new workflows.

51 We visualize reasons for decisions.

52 We have established "Best Practice" standards.

53 Decisions are made where the competence is the highest.

58 Requirements for competence, qualifications, and responsibilities for all jobs are visualized.

60 The system supports coordination across HQ and subsidiaries (including transfer pricing).

64 There are consequences for not delivering time recording one time per week.

67 We get better at finishing projects.

B3.2 The system supports optimal project implementation

2 Reliable data that are validated before they come into the system.

7 The system is integrated with other information systems so that the information is coordinated across these systems.

8 Visible utilization of resources (billable time versus time spent).

9 The system must make it easier for the users to do planning.

17 Utilization of the system gives a measurable economic benefit.

19 We get an overview of where we lose and make money so that we get an improved business focus.

22 The invoice foundation appears significantly faster.

23 Increased maneuverability.

24 Flexible reporting capabilities that show relevant information.

25 Each employee has an overview of the tasks that the employee is allocated to and used on.

26 Be able to identify potential conflicts and deviations early in the project.

33 Bottlenecks are visible.

34 Internal and external plans can be maintained in the same place.

37 The system is easily adaptable to new methods at WCAT.

39 It will be visible if a task is behind or ahead of schedule.

45 The system is proactive—it reminds the user about activity that must be initiated.

47 No excuse for not being proactive and for not taking initiative.

50 We can assess the impact of different projects and project portfolio scenarios.

56 The system contains only one truth.

57 We get fast and useful final costing of projects.

59 The system can highlight the vulnerability relative to essential staffing.

65 You can register all kinds of time spent in the system.

B3.3 Conformity between our delivered services and the customer expectations

10 Controlled project process with fewer surprises, higher predictability.

16 The planned project times are respected.

20 We can at any time inform the customer about the status of the costs incurred in the customer's projects.

23 Increased maneuverability.

28 Higher customer satisfaction because we deliver what is agreed on time.

29 Visible customer deliveries and consequences of the customer's failure to comply with agreed conditions.

31 The system provides a better basis for guiding the customer to an optimal process.

36 Better time estimation for proposals (standard time).

40 It becomes apparent who must be notified when there are deviations from a plan.

42 All involved stakeholders report problem situations in the system with the assurance that the situations are treated in time.

48 We must be able to detect and measure the quality of each process (e.g., agreed with the customer).

B3.4 Applied and accepted management tool

4 The sales team uses the system for capacity calculation and budgeting of potential projects.

11 Increased reusability of collected data and experience.

12 Better differentiated classification of time recording.

15 The system is embedded in the organization—also in 2 and 3 and *n* years.

35 The projects must be started and built uniformly when they are of the same type.

38 The plans are available on the intranet.

41 The system's benefits are visualized and communicated.

54 There is education internally in the system even after it has come into operation.

55 We start up with a solution that quickly provides visible benefits and is widely used.

61 The system data are protected against unauthorized access.

62 It can be seen who has recorded system data.

63 We get a proven method per type of project.

66 The system demands very limited operational intervention.

68 Accepted tool for leadership and management.

B3.5 The solution contributes measurably to increased profitability

8 Visible utilization of resources (billable time versus time spent).

11 Increased reusability of collected data and experience.

17 Utilization of the system gives a measurable economic benefit.

19 We get an overview of where we lose and make money so that we get an improved business focus.

22 The invoice foundation appears significantly faster.

36 Better time estimation for proposals (standard time).

55 We start up with a solution that quickly provides visible benefits and is widely used.

57 We get fast and useful final costing of projects.

B3.6 Conformity between LI services and WCAT expectations

69 LI meets agreed deadlines.

70 LI coaches throughout the implementation.

71 WCAT is responsible for project management—LI ensures an efficient process.

72 WCAT are satisfied with LI services.

B4. PQA MATRIX

Critical Success Factors:							EXECUTION VALUE	RESPONSIBLE
6) Conformity between LI services and WCAT expectations								
5) The solution contributes measurably to increased profitability								
4) Applied and accepted management tool								
3) Conformity between our delivered services and the customer expectations								
2) The system supports optimal project implementation								
1) Effective project culture								
Activities:	**1**	**2**	**3**	**4**	**5**	**6**		
1) Establish project team	*	*		*		*	1	MI
2) Establish project office	*	*	*	*			0	HK
3) Define and document the essential project types		*	*		*		2	ML
4) Define and document the optimal project management— responsibilities, competences, rights, duties		*	*	*	*	*	2	JO
5) Clarify and implement Phase 1				*	*	*	0	CT
6) Develop user guide for WCAT use of the system—parameters, security, procedures, standards		*		*	*	*	2	JO
7) Define metrics for earnings				*	*		0	LN
8)Clarify course scope, content, and audience—conduct them	*			*		*	0	LM
9)Inform the entire organization continuously about opportunities, goals, effects, and requirements	*			*	*		0	LM
10)Install the system with established standards and access to desired data and functionality for all users		*		*		*	0	CT
11) Acceptance testing of Phase 1	*	*	*	*	*	*	0	LI

B5. ACTIVITY DESCRIPTION EXAMPLES

B5.1 Define and Document Essential Project Types

Activity Description	3) Define and document essential project types
Prepared by ML	Date: 06.04.12
Delimitation	This activity does not include the following: • The specific project model and the contents of the activities in the individual departments • The organizing of the projects and the roles in the WCAT (project)-organization
Products	A document that describes the essential project types with clearly defined phases, when to change phase, and the distribution of responsibilities. During this we should determine the interfaces and dependencies between the various functional areas.
Purpose	The purpose of describing the essential project types is to define and document these in a structured way, and from this to: • Be able to prioritize based on WCAT business objectives • At any time, be able to see in which phase a project is and who is responsible • Ensure that the individual employee has an overall view of the phases in essential project types • Ensure a structured collection of experience-figures for each project type • Ensure that the descriptions form the basis of the implementation in the COTS
Responsible	JO
Other resources	CT; ML Qualifications: Thorough understanding of WCAT business processes (service, workflow and products)

Sub-activities	Description of Tasks	Resource Requirement
	1) Identification of the essential project types.	JO, ML (4 hours)
	2) To carry out the analyses and describe the individual project types a group (person) per project type is established. It is the responsibility of the individual group to: • Describe the existing project workflow • Identify the action areas/problems • Determine and describe the ideal project workflow (phases, when to change phase, etc.) • Describe the distribution of responsibilities for the project workflow	Expected time consumption per group: 2 weeks
	3) Configure COTS for the essential project types.	CT, LI, ML, JO Expected time consumption: 4 weeks
Time frame	August–October 12	
Risks	If this project is *not* given top priority by the management, the resources will disappear from the project. A strong project team must be set up to ensure its visibility in the organization. Efficient project culture does *not* arise of itself. COTS is *not* able to support our description of the project types. The system is *not* easily adapted to new project types.	
Dependency on other activities	This activity can be started independently of other activities. Close coordination with: 3) Define and document effective project management is necessary.	

B5.2 Define and Document Optimal Project Management

Activity Description	4) Define and document optimal project management	
Prepared by JO	Date: 08.19.12	
Delimitation	The definition and documentation must reflect the complexity of the project types we have in WCAT and also the resulting demands on managing and reporting ("We do not build spacecrafts")	
Products	Clearly defined project workflows and established routines on workflow management	
Purpose	To create a common understanding of and experience in how projects should be managed so that our customers see us as professionals who keep their word and so that we continuously improve our efficiency and quality based on the experience gained	
Responsible	JO	
Other resources	Departmental managers or others who are responsible in relation to our project types, Project Team, and Project Office	
Sub-activities	**Description of Tasks**	**Resource Requirements**
1.1.1	1) Define the required description for all project types to ensure consistency in views and results	Steering Group and Project Team
1.1.1	2) Describe the workflow of the individual types based on start-up, implementation and completion criteria in accordance with the agreed template	Project Team together with the persons responsible for a workflow
1.1.1	3) Describe roles in a workflow (participant types, departments involved)	Project Team and Steering Group
1.1.1	4) Describe distribution of responsibilities including handover requirements and the related documentation (the baton), which must exist between processes	Project Team together with the persons responsible for a workflow

Sub-activities	Description of Tasks	Resource Requirements
1.1.1	5) Determine QA in a workflow to prevent errors	Project Team together with the departmental managers and Steering Group
1.1.1	6) Determine measuring points in a workflow as regards time consumption to be able to: • Estimate time for cost estimation • Estimate time for planning • Measure earnings	Project Team together with the departmental managers, LN, and Steering Group
1.1.1	7) Make cost accounting	Project Team and Steering Group
1.1.1	8) Implementation form: How do we define implementation What signifies that a project has been handed over?	Project Team together with departmental managers
1.1.1	9) Formalization of the cooperation with the customer before, during, and after	Project Team and Steering Group
1.1.1	10) Continuous follow-up (what can be made more efficient in a similar workflow in the future), especially after the first implementation	Project Team, Project Office, and Steering Group
1.1.1	11) Success criteria for a project type should be determined (did the customer get what we promised)	Project Team with Sales
1.1.1	12) Handover of the project to Support	Project Team
Time Frame	Effective time: 1 week's general preparation and 3 weeks per project type (workflow). Additionally 1 week per workflow for finalization of description.	
Risks	The project cannot be given the priority that was intended The selected resources cannot participate at the right time The dependent activities are delayed The above-mentioned time estimate is wrong	

Dependency on other activities	Predecessors: 3) Define and document the essential project types 1) Establish Project Team 2) Establish Project Office 8) Internal courses must be held 11) Tools must be available

B5.3 Clarify and Implement Phase 1

Activity Description	**5) Clarify and Implement Phase 1**	
Prepared by CT Revised by LI	Date: 06.17.12 Date: 08.24.12	
Delimitation	Install the system in accordance with the specifications in the contract between ST and LI	
Products	COTS installed to be used for time registration and project and production planning in ST (DK, UK, US) Define time frame and schedule for introduction of the individual elements in COTS After the implementation, the system will be handed over from the project team to an operation organization	
Purpose	Ensure that the contractual obligations between ST and LI work as expected Ensure technical functionality of delivered COTS	
Responsible	CT	
Other resources	LI, ST UK, ST US, IT	
Sub-activities	**Description of Tasks**	**Resource Requirements**
	Define exact content of Phase 1	CT 40 hours LI 40 hours
	Prepare extra functionality: Integration COTS and SAP Production planning Copying of project	CT 120 hours LI max 105 hours LI max 180 hours LI max 95 hours
	Test: Connection to ST, UK, and US	CT, ST, UK, and US 20 hours LI max 8 hours

	Install ST in DK, UK, US: Time registration Project management Production management	CT 40 hours ST UK 20 hours ST US 20 hours	
	Handing over of running system to the operation organization	CT 20 hours IT 40 hours	
Time frame	Extra functionality: Integration COTS and SAP Production planning Copying of project	September 30, 2012 December 15, 2012 January 31, 2013	
	Installation	ST DK	ST UK	ST US
	Time registration	September 1, 2012	January–May 2013	June– September 2013
	Project management	Mid September 2012 (old projects)		
	Production management	December 2012		
	The handing over to the operation organization must take place continuously as the system elements are put into operation.			
Risks	Delayed specification of extra functionality Delayed definition of the contents of training courses Delayed definition of project types Delayed definition of effective project management			
Dependency on other activities	The installation of the system can only take place based on the results/part results of the following activities (predecessors): 6) Develop user guide for WCAT use of the system 8) Clarify courses scope, content, and audience and conduct them			

B5.4 Develop User Guide for WCAT Use of the System

Activity Description	6) Develop user guide for WCAT use of the system	
Prepared by JO	Date: 08.19.12	
Delimitation	The handbook comprises only the necessary information. This means that it should not be thicker than absolutely necessary and it must be easily comprehensible.	
Products	A handbook that describes the routines and disciplines which must be maintained in relation to our project types.	
Purpose	That an employee who plays a part in a project type can see: • What should be done in the course of the project • How it should be done • What is expected of him or her in the course of the project • What he or she may expect from other participants in the project That any new employee, as part of an introduction program, can obtain the above insight.	
Responsible	JO	
Other resources	The project team and departmental managers in connection with revisions to ensure that the everyday life is reflected.	
Sub-activities	**Description of Tasks**	**Resource Requirement JEO + Project Team**

	1. Define the expectations to the handbook
	   • Easy to use including good examples
	   • Easy access to information
	   • Always up-to-date
	   • A picture is worth a thousand words
	2. Make the skeleton of the book
	3. Make a good table of contents
	4. Produce the contents
	5. Proofread
	6. Finalization
	7. Establish an updating routine
		Departmental managers

Time frame	Must be prepared concurrently with other processes toward the implementation (training before implementation).
Risks	The project cannot obtain the intended priority The selected resources cannot participate at the right time The dependent activities are delayed The above time estimate is too optimistic at the present time
Dependency on other activities	The following activities must have been completed: 1) Establish Project Team 2) Establish Project Office 3) Define project types 4) Define effective project management The handbook may be prepared concurrently with, for instance, the finalization of project types.

Appendix C: Delivery of Consultative Services Framework Agreement between Bank and Solution Provider Contract No. 999

C1. GENERAL

This Framework Agreement between the *Bank* and the *Solution Provider* concerns delivery of IT consultative services.

Apart from this Document, the Framework Agreement consists of the following Attachments:

Attachment A:	Consultative Services Agreement
Attachment B:	Statement of Confidentiality

C2. OBJECT AND SCOPE OF THE AGREEMENT

The object of this Framework Agreement shall be to set out the terms and conditions for the consultative service rendered by the *Solution Provider* to the *Bank*. The terms and conditions laid down herein shall apply to any consultative service rendered by the *Solution Provider* on behalf of the *Bank* notwithstanding whether an agreement has been entered into in writing. However, as for term and prices the Agreement shall not be valid until after proper signing by both parties.

The detailed character of the consultative assistance, time consumption, and remuneration therefore shall be agreed and described in Attachments A and B, which shall be drafted in duplicate original and signed by both parties.

In the event of discrepancy between the Consultative Services agreement and this Framework Agreement, this Framework Agreement shall prevail unless a deviation from the Framework Agreement has explicitly been described in the Consultative Services agreement.

C3. DEFINITION OF THE PARTIES

This Agreement shall apply to the *Bank* and its Affiliated Companies.
"Affiliated Companies" of the *Bank* means any legal entity:

- Directly owning or controlling the *Bank*
- Under the same direct or indirect ownership or control as the *Bank*, or
- Directly or indirectly controlled by the *Bank*

for so long as such ownership or control lasts. Ownership or control shall exist through direct or indirect ownership of more than fifty percent of the nominal value of the issued equity share capital or of more than fifty percent of the shares entitling the holders to vote for the election of directors or persons performing similar functions or right by any other means to elect or appoint directors or persons who collectively can exercise such control.

C4. THE *BANK'S* OBLIGATIONS

The *Bank* shall be obliged to make all of its relevant facilities available to the consultant, including office facilities, machine time on the *Bank's* computer platform and assistance from agreed to available *Bank* employees to the extent necessary to ensure that the consultant can perform his or her work based on the consultant's agreed competence and skill.

C5. STAFF

The *Solution Provider* shall only make consultants available, who have been approved by the *Bank* to possess the qualifications required by the *Bank*.

If the *Bank* so wishes, the *Solution Provider* shall provide the *Bank* with registrations of time spent for the consultative services rendered.

The *Solution Provider* agrees that the *Bank* is entitled to request that a consultant be substituted upon submission of reasonable arguments, including illness. In such event, the *Solution Provider* shall provide another consultant who shall possess the professional qualifications. The change of consultant including the time spent to hand over the contracted service shall take place without any costs or delivery time implications for the *Bank*.

The *Bank* shall be obliged not actively to employ or in any other way attach the *Solution Provider's* consultants to its business without the prior approval of the *Solution Provider* within the terms of this Agreement.

The *Bank* shall be entitled to terminate this Framework Agreement forthwith if the *Solution Provider* without written approval from the *Bank* employs or in any other way attaches an employee with the *Bank* or an employee who was employed with the *Bank* during the last 12 months.

In the event that it becomes necessary to use a sub-solution provider, the sub-solution provider must co-sign the Attachments and the Framework Agreement. For the purposes of this Framework Agreement, sub-solution provider shall mean all persons and/or companies who/which the *Solution Provider* directly or indirectly uses to perform its obligations and who/which are not employed with the *Solution Provider*.

The *Bank* shall be entitled not to use a sub-supplier, including named persons employed with the *Solution Provider* or with a sub-supplier.

Neither the *Solution Provider's* use of sub-suppliers nor the *Bank's* rejection to use sub-suppliers/named persons shall in any way change the *Solution Provider's* obligations as set out in this Framework Agreement.

C6. REMUNERATION

The *Solution Provider* shall invoice the *Bank* as specified in Attachment A. Terms of payment, invoice date + 30 days.

C7. WORKING HOURS

The working hours shall be all business days within ordinary hours. In case the *Solution Provider* needs to work outside these hours, access to the premises must be agreed in writing with the security officer.

C8. TRANSPORT (WITHIN COUNTRY)

Costs for daily transportation from the location of the *Solution Provider* to the location of the *Bank* shall be paid by the *Solution Provider*.

C9. TRAVELING AND STAYING (OUTSIDE COUNTRY)

Costs for traveling and staying outside the country upon written instructions from the *Bank* shall be directly payable by the *Bank*.

C10. INTELLECTUAL PROPERTY RIGHTS

All programs, details of specifications, screen structures, plans, user guides, and documentation, including source texts developed under this Agreement shall be the property of the *Bank* so that all intellectual property rights and other rights shall belong to the *Bank* unless otherwise agreed in Attachment A to this Framework Agreement.

Unless explicitly otherwise agreed or informed, the *Solution Provider* shall guarantee that the documentation and other results provided by the *Solution Provider* and delivered to the *Bank* do not infringe third-party rights.

C11. STANDARDS AND SECURITY RULES

The *Solution Provider* shall be obliged to use and observe the standards and rules for IT system development applicable to the *Bank*, the *Solution Provider's* own company rules of ethics, and the *Bank's* security rules.

C12. CONFIDENTIALITY

The parties and their employees shall be subject to an unconditional duty of silence as for any knowledge of, information on, or documentation of the other party's internal matters, plans, products, finances, clientele, etc., which the other party or its employees have learned through the cooperation.

The duty of silence is irrevocable.

This is confirmed by each of the consultants through the signing of the *Bank's* Statement of Confidentiality (Attachment B).

The *Solution Provider* shall not use the *Bank* as reference without the prior written approval of the *Bank*. The *Solution Provider* shall not send out public information on this Framework Agreement or publicize parts of or the entire contents of this Framework Agreement without the prior written approval of the *Bank*.

C13. FORCE MAJEURE

Under this Framework Agreement, neither the *Bank* nor the *Solution Provider* shall be deemed to be in breach of their obligations in relation to the other party if such breach is solely due to external events on which the parties have no influence themselves.

Force majeure relating to delay shall not be claimed for a number of hours in excess of the number of hours for which the force majeure situation lasted. Force majeure shall only be claimed if the party in question has informed the other party thereof in writing no later than 4 working days after the occurrence of the force majeure situation.

However, the party not affected by the force majeure situation shall be entitled to cancel the Agreement if the agreed time of delivery is exceeded by more than 10 days due to force majeure. In such event, both parties shall as soon as possible return everything which they have received from the other party and no further claims shall exist between the parties.

C14. TERM AND TERMINATION

This Framework Agreement shall be in force for a period of 1 year as from the commencement date. The Agreement shall be up for re-negotiation no later than 3 months prior to expiry.

This Framework Agreement may be terminated by either party giving three months' notice in writing until the first day of a month.

However, this Framework Agreement shall be non-terminable and without limitation in time as for the rules on confidentiality, rights, governing law, and arbitration.

C15. GOVERNING LAW AND ARBITRATION

This Framework Agreement shall be governed by Country law.

Disputes arising between the parties and which cannot be settled amicably shall finally and with binding effect be settled through arbitration in accordance with "the Rules of Procedure" for Country Arbitration. The award rendered by the arbitration tribunal shall be based on Country law. As for the question of legal costs, the arbitration tribunal's award shall be based on applicable rules on legal costs for legal proceedings.

This Framework Agreement has been signed in duplicate original and each party shall receive one original hereof.

Date: Date:

_____   _____
For and on behalf of the *Bank*   For and on behalf of the *Solution Provider*

Appendix D: Consultative Services Agreement under Delivery of Consultative Services Framework Agreement 999 between Bank and Solution Provider

D1. REFERENCE TO THE FRAMEWORK AGREEMENT

Contract number 999

D2. DESCRIPTION OF THE TASK AND OBJECTIVES

Name of work package and reference to work package document. Further specifications as agreed to.

D3. CONSULTANCY SERVICE

Delivery of agreed work package on fixed price fixed delivery date. Special conditions deviating from the Framework Agreement.

D4. CONTACT PERSONS—*SOLUTION PROVIDER* AND *BANK*

The contact person or a named backup person must be available to talk about any issue with a delay of no more than one hour on any working day. It is the Solution Provider's responsibility to inform the named Bank contact person about who is to be used for contact person, when this person deviates from the contact person in this agreement.

D5. START/END

Start is upon signature of this Consultative Services Agreement. End is at latest ____/____/____.

D6. PLACE OF WORK

The bank dedicated premises.

D7. PRICE

The price is EUR ____ ____ ____.
The price is exclusive of TVA, travel, and accommodation.
Travel and accommodation is based on documented reasonable costs and must be approved in advance for consultants with residence outside Country.
__% of the price can be invoiced after 25 working days of satisfactory progress as judged by the Bank.

Up to 100% of the price can be invoiced after satisfactory delivery of the work package solution signed off by the bank IT Director.

D8. NOTICE OF TERMINATION

In case of termination without delivery of the agreed work package solution the Solution Provider is not entitled to any form of remuneration.

For each working day of delay after the end date agreed in §5 the price under §7 is decreased by 1%.

A delay of more than 20 working days after the end date agreed in §5 terminates automatically this agreement and the Solution Provider is not entitled to any form of remuneration.

D9. SIGNATURES OF SOLUTION PROVIDER/BANK

Date: Date:

_____   _____

For and on behalf of the *Bank*   For and on behalf of the *Solution Provider*

Appendix E: Introduction to Participants in Information Requirements Study

E1. SCOPE

An Information Requirements Study (IRS) uncovers the required information, information systems, and procedures needed for the execution of business functions.

IRS is performed whenever the organization needs important changes and improvements in the way it is doing business for many different reasons:

- Legal changes
- Market condition
- New products
- Merging with other organizations
- Splitting up activities into separate organizations

IRS is performed at three levels of the analyzed organization:

1. **Section** level, which includes employees and managers from the operational areas such as Stock Control, Accounting, Sales Order Handling, Production Planning, Client Service, Back Office, Trade Desk, etc. dealing directly with production, acquisition, sales, and delivery of the corporate products. The sections are defined especially for IRS to ensure that all business functions are covered by the study
2. **Departmental Management** level, which includes managers who are responsible for the planning of coherent operational activity, such as financial management, production management, logistics, quality, portfolio management, compliance, etc. Also on this level it is ensured that the managers cover all business functions.
3. **Executive Management** level, which includes leaders who are responsible for the organization's policies and the strategic initiatives.

The IRS process, including selection of interview/review participants, is coordinated by a cross-organizational IRS Reference group and an IRS Workgroup. The IRS Workgroup with a facilitator (most often external to the studied organization) and an internal IRS project manager supported by a member of the IRS Reference group perform all studies/interviews on all IRS levels.

The resulting documentation is produced by the IRS facilitator in collaboration with the interview participants (documentation owners), the IRS project manager, and the IRS Reference group.

E1.1 Product

The IRS result is presented as a set of reports of which the concluding IRS consolidated report is a requirements specification for one or more solution components. A solution in this context is a complex combination of:

- Organizational change requirements (business roles, functions, procedures)
- Requirements to IT-based information systems
- IT requirements (technology, security, safety, reliability, integrate-ability, availability)

Each IRS result explains:

- Why the suggested solution is required
- Why an existing solution can been used
- Why an existing solution has been made obsolete or needs improvement

In this way IRS creates a commonly accepted foundation for future programs and projects concerned with planning, development, implementation, and governance of corporate information system solutions.

The IRS consolidation process uses Object Lifecycle Analysis and documents types such as:

- Cross-reference tables
- Entity Relationship Diagrams (ERD)
- Data-flow diagrams (DFD)
- Create, Read, Update, and Delete (CRUD) tables

The document types are tools for data and process normalization.

The normalization of data and processes ensures solution integrity and consistency.

E1.2 Objective

The requirement for information is documented in such a way that it visualizes how the users of the future solutions can optimize their working processes and their roles and behavior on operational and management levels.

E1.3 IRS Terminology

E1.3.1 Objects and Entities

We do not regard these as different for the purpose of IRS. Entities is a practical term from relational database terminology which fits the normalization need well, while objects is a term from object-oriented system development that fits our need for open systems well.

You don't need to regard them as different.

E1.3.2 Interviews

IRS interview reports are created based on interviews or workshops.

Each IRS interview is performed with participation of:

- Interviewers are the IRS facilitator and the IRS project manager (the IRS Workgroup).
- A key person from the IRS Reference group participates if possible in all interviews.
- Interviewees are selected employees and managers from the business functions covered by the interview, that is, the best available experience and knowledge motivated for change.

The selected persons prepare themselves for the interview by collecting pertinent documentation used during their business processes as results or guidelines (e.g., Quality System documentation, forms, etc.).

The interviewers and the IRS Reference group key person support the interviewed employees and managers to describe:

- Responsibilities (Why)
- Work processes (What)

- Information requirements (available or not available, needing or not needing improvement)
- Communication with internal and external organizations

The final report is approved and owned by the interviewees and it can only be updated by the interviewees.

E1.3.3 Sections

A section is a logical organizational unit on an operational level that is specific to the conduction of the IRS. The section is responsible for the quality of a set of results that are used inside the section or by other sections in the pursuit of delivery of high quality results to the users and buyers of the corporate products.

Employees in a logical section are dealing with related tasks in order to deliver complete and consistent results, which comply with the overall policy and strategy of the corporation.

Normally 1 to 4 employees from a section possess all knowledge and experience needed.

Managers and employees in a section are normally concerned with the following types of information and functional relationships:

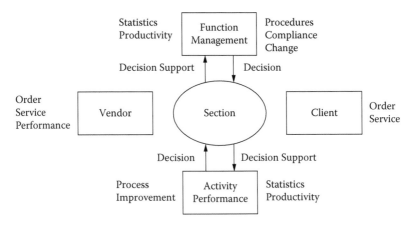

E1.3.4 Departments

Managers on a departmental level use their information for decision support in short- and long-term decision-making processes.

Departmental managers are normally concerned with the following types of information and functional relationships:

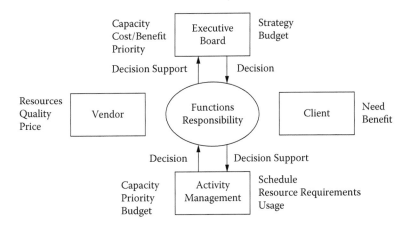

On this basis, the Departmental IRS report gets a structure (see below), which reflects this assumption.

E1.3.5 Executive Board

Executive board members are normally concerned with the following types of information and functional relationships:

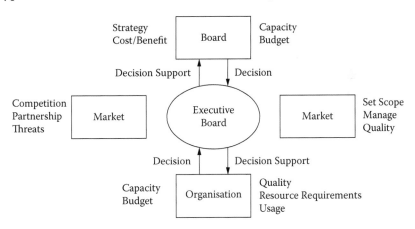

On this basis, the Executive Board IRS report gets a structure (see below), which reflects this assumption.

E1.3.6 Consolidation Workshop and Consolidated IRS Report

The IRS consolidation is finally performed by the IRS Reference group and the IRS Workgroup who produces the consolidated IRS report.

When the IRS process is ready for consolidation after approval of all interview reports, new members of the IRS Reference group might be required in order to ensure state-of-the-art and business compliant recommendations prior to approval by the sponsor (the Strategy Governance group).

The IRS consolidation workshop establishes a common terminology, which is used in the IRS Consolidated Report. This report describes the structure and contents of required information, and it describes the information systems, the technology, and the business functions, which are required in order to ensure availability, reliability, validity, security, and consistency of this information.

E1.4 Procedure

All participants receive this introduction and participate in one or more kick-off presentations (3 hours' duration), where their roles and responsibilities are outlined and agreed to, and the expected results are presented.

There is one interview and 1 to 2 reviews for each section and each manager/management group.

It is recommended that participants in interviews prepare themselves by listing their current and future requirements for information in support of their tasks and decisions in connection with:

- Planning and controlling work procedures
- Reporting requirements from/to internal and external sources

It is recommended that the participants bring documentation and other examples covering:

- Forms and documents used
- Reports, minutes, etc.

E1.5 Areas of Responsibility

The IRS facilitator is responsible for ensuring that all participants are properly informed about the IRS process, their role, and their IRS responsibilities.

The IRS facilitator writes the first draft interview report which shows the interviewee how the report should be structured. This draft cannot be expected to have the content quality that can be accepted and owned by the interviewee without a thorough review and approval process.

The IRS participants are expected to talk openly and describe their requirements and wishes as they interpret them in their own language during the interview. The interviewers will have to understand this language in order to be able to document the requirements.

The interviewed participants are fully responsible for the final content quality of their report. The interviewees own their report and are invited to improve this report as required in the future.

The IRS Reference group supports the IRS facilitator and the IRS project manager. The IRS facilitator and the IRS project manager keep the interview participants informed about the IRS progress so the participants understand their role at any point in time:

- Interview preparation
- Report review
- Report writing
- Report sign-off
- Report distribution

E1.6 Interview Content

The interviews visualize the participants' information requirements in order to solve their daily tasks in compliance with their responsibilities and authority by answering the following questions:

- What are their tasks and the objectives of these tasks, and why?
- What is the importance of these tasks as seen from the other parts of the organization, and why?
- What information is used for what, and why?
- What new or enhanced information will make it easier to handle the tasks, and why?
- What information used originates from others, for example, from management, other sections, other departments, external partners, and why?
- What information used is archived and maintained, and why?
- What information used is delivered to others, for example, to management, other sections, other departments, external partners, and why?
- How well do the existing solutions (with or without IT system) support the way we solve our tasks, and why?

- How can integration of solutions enhance communication and improve information quality, and why?
- Suggested solutions, and why?

It is recommended that the interviewees bring examples of forms and reports in support of explanation, issues, and recommendations.

E1.7 Interview Result

Each interview is documented in a report with a predefined format. It describes the workflows, the required decisions, and the information requirements in the terminology of the interviewees.

At the same time, the report has a structure and a content, which makes it an agreed common framework of reference for the interviewees and the internal and external service providers who must deliver efficient solutions in support of the interviewees later on.

The approved interview report belongs to the interviewees and is distributed to relevant management levels when it has been approved by the interviewees.

E1.7.1 Section Report

A section report has the following defined structure and content:

1. **Scope, product, and purpose**
   Scope describes the extent of the section's responsibilities and functions, including who the principal "suppliers" and "customers" are.
   Product describes the services and physical products that the section's activity results in.
   Purpose describes the expected and desired quality of the section's products, including their ability to fulfill the section's and the company's needs. The purpose should be related to the outside world need that is covered by the section and the section's own quality requirements.
2. Available information and its usage
   For each of the section's responsibilities/main functions it is described how the required information is obtained, used, and released to/ from other sections and other organizations.
3. Required information, which is currently not available

It is described how additional or better information can improve the efficiency of responsibilities/main functions of the section, including the quality of products supplied to other sections or company stakeholders, such as owners, customers, and suppliers.

All the expressed wishes must be justified; WHY must be answered accurately with indication of the benefits obtained.

4. The value of the required information

   The specific advantages that can be obtained from availability of the required information are listed, including their value to the organization if it can be estimated.

5. Information Flows

   Examples shown:

Information received by section	*from section/others*
Application form	Potential tenant
Information maintained by section	
Tenancy agreement	
Information delivered from section	*to section/others*
Tenancy agreement	Tenant

6. Suggested improvements to available IT and functional solutions

   Wishes for better procedures, both manual and IT-based, that could support the generation and maintenance of better information and communication.

   Also inconveniences in existing information systems are highlighted. A detailed review of existing solutions is not sought unless it is a prerequisite to understanding why the solution is inappropriate.

   All the expressed wishes must be justified; WHY must be answered accurately with indication of the benefits obtained.

7. Suggested integration of systems

   The need for the establishment of communication where it does not exist and for improvements to existing communication where there are problems of quality, timing, or information content.

   Highlight problems associated with duplication of registration or lack of access to consistent information that currently is spread over disconnected "islands."

   All the expressed wishes must be justified; WHY must be answered accurately with indication of the benefits obtained.

8. Expected value of improvements

   The specific advantages of better information systems and business procedures and integration are listed, including their value for the company if it can be estimated. The value is allocated on benefits and savings.

9. Suggested (new) IT solutions
Here is free play for the section's interview participants and interviewers; your imagination can be used to the best effect. However, it must be remembered that all the expressed wishes must be justified; WHY must be answered accurately with indication of the benefits obtained.

10. The following types of annexes can be attached:

- Descriptions of all identified objects attributes and usage
- DFD for the section
- ERD

E1.7.2 Department Report

1. Scope, Products, and Purpose
Scope describes the extent of the department's responsibilities and functions, including who the principal "suppliers" and "customers" are. **Product** describes the services and physical products that the department's activity results in. This product can be described more qualitatively than the related sections; it is often a framework for these.
Purpose describes the expected and desired quality of the department's products, including their ability to fulfill the section's and the company's needs.

2. Wishes for improved information
Need for information related to the essential decisions that are the primary product, for which the department level is responsible.

3. Potential improvements in decision support
It is specified here how improved information creates better overview, more consistent information, and more valid information, which together reduces uncertainty and risk in the department's decision making and budgeting/planning of business operations and Strategic Initiatives.

4. Expected benefits from improvements
The specific advantages that can be obtained from availability of the improved information are listed, including their value to the organization even if the advantage is intangible.

E1.7.3 Executive Board Report

1. The company's overall business objectives
   - Scope (business area/industry)
   - Product (market)
   - Purpose (quality, market share, growth)
2. Wishes for improved information
   Need for information related to the essential decisions that are the primary product, for which the executive level is responsible.
3. Potential improvements in decision support
   It is specified here how improved information creates better overview, more consistent information, and more valid information for the governance of current business and strategic initiatives.
4. Expected benefits from improvements
   The specific advantages that can be obtained from availability of the improved information.

E1.8 Quality Assurance and Review

The interviewees will normally receive the resulting IRS section or management report within 2 days after the interview.

The interviewees have at least two days to review and suggest corrections to the report before they participate in one or more review meetings, where the report is finally written and approved. The IRS Workgroup and the IRS Reference group members support the interviewees in their review process.

The final report whether written by the IRS facilitator or the interviewees themselves is approved by the interviewees and is regarded as property of the interviewees that are listed on the section and management report front page.

E1.9 Conclusion and Recommendations

Consolidation of the different interview reports into the IRS Consolidated Report is solved using Object Lifecycle Analysis (OLA), which is a method with its own document standard.

OLA uses the different objects that were identified during the interviews:

- Action objects (Sale, Purchase, Fee, Corporate actions, Claims, etc.)
- Structure objects (Client, Location, Security, Dealer, Custodian, etc.)

OLA ensures complete requirements in the form of business processes, action information objects, and structure information objects.

The IRS Reference group supported by the IRS project manager and the IRS facilitator participate in the OLA workshop.

The IRS consolidation workshop establishes a common terminology, which is used in the IRS Consolidated Report. This terminology is documented in the Vocabulary, which contains the following:

- Term (word or abbreviation)
- Definition
- Usage example

The IRS Consolidated Report describes the structure and contents of required information, and it describes the information systems, the technology, and the business functions and procedures, which are required in order to ensure availability, reliability, validity, security, and consistency of this information in the conduct of the corporate business covered by the IRS.

The solution requirements documented in the IRS Consolidated report are in classic text, but graphics elements are used as well:

- DFD
- Workflows
- ERD
- Input/output matrices
- Role/Responsibility diagrams

The OLA method is also used for solution design, where it is supplemented with user interface design based on technological opportunities and other documents standards such as described in Chapter 5. Detailed solution design is not in the scope of IRS.

E1.10 Annexes to the IRS Introduction

The IRS interview participants before the interview must be informed about the IRS project objectives, the IRS Scope. It is therefore important that the following material from Process Quality Assurance (PQA) in the IRS Reference group is attached to this introduction to the interview participants:

- IRS Reference group visions
- Critical Success Factors
- PQA Matrix
- Guide to Object Description
- Forms for Object Description and vocabulary elements

In the IRS introduction meeting, a copy of the PowerPoint presentation used should be available.

Appendix F: FOSIS Information Requirements Study OLA Consolidated Report Extract with Summary and Conclusion 2013

F1. SUMMARY

FOSIS must support all defense healthcare tasks:

- Consultation and treatment on infirmaries with and without dentist clinic, ships, and field hospitals
- Aviation and diver medical consultations
- Patient logistics in peace, crisis, and war
- Material logistics and maintenance in peace, crisis, and war
- Personnel management
- Quality Management and Administration

FOSIS will comprise the following modules to be implemented in the shown sequence:

1. Consultation
   Treatment
2. Patient logistics
3. Personnel planning
4. Procurement
   Inventory management
   Material maintenance
   Inspection
5. Development

As the modules are defined here they support all healthcare tasks in the defense with at least the following benefits:

- Benefit description

The value of the benefits has been pessimistically estimated to at least €4 million per year through:

- Better utilization of resources
- Healthier personnel
- Improved quality of offered healthcare services

We estimate that the full value of benefits can be earned over a 3-year implementation period.

First year cost based on experience with the full implementation of a COTS-based solution across all defense barracks is estimated to be €3 million.

If the defense decides to base the solution on the integrated defense COTS application that is not prepared for a FOSIS solution, the first year costs and the annual operation costs will be considerably higher.

Annual operation costs are estimated at €1 million per year.

The suggested FOSIS solution with requirements to integration, technology, and organization fulfills all Critical Success Factors from the PQA that initiated this IRS.

The estimates do not take into account adaptation to international defense standards except for the WHO terminology standards and the already established NATO standards. These standards must be adhered to in FOSIS as they become available.

The future COTS vendor for FOSIS must guarantee compliance of functions, data, and reporting terminology and classification concerned with national and international healthcare standards at delivery and by delivery of adequate adaptation and new versions in the future.

F2. THE FOSIS TASK

Replace all current systems and improve the services provided by DHS.

The primary wish is to get implemented an electronic health journal that follows all personnel wherever they are in peace, crisis, and war.

The health journal must function irrespective of geographical placement.

F2.1 Introduction

DHS (Defense Healthcare Service) requires new Information Systems.

F2.2 The DHS Situation and Initial Conditions

DHS employs 999 persons distributed as doctors, dentists, nurses, assistant nurses, and administrative personnel.

F2.2.1 DHS Tasks

Primary DHS tasks comprise:

- Prevention and treatment of military personnel health problems. In peace time, the military personnel is conducted to the civil healthcare system for treatment.
- Military healthcare training of an important number of doctors and healthcare personnel that can be called upon under crisis, catastrophes, and war.
- Support of civil institutions concerned with helicopter-based surveillance and sea rescue tasks including development of special equipment to this end.

F2.2.2 DHS Current IT-Based Systems

F2.3 The IRS Method Used

The method used to prepare this requirements specification is IRS.

F2.3.1 IRS

IRS comprises the following activities:

- Establishment of study scenario
- Definition of sections (virtual organization that encompasses all DHS tasks)
- Selection of participants (Facilitator, Workgroup, Reference Group, Section interview participants, etc.)
- IRS Introduction to participants
- Interviews with section participants

- Documentation of task-based sectional information requirements
- Review and approval of section IRS reports
- Management level interviews and IRS management report writing, review, and approval
- Consolidation workshop with IRS Workgroup and IRS Reference group to write this report
- IRS result sign off by sponsor

The IRS has been done on three organizational levels of DHS:

1. Sections with managers from:
   - Infirmaries with and without dental clinic
   - Air force stations (aviation medicine)
   - Fleet stations (diver medicine)
   - Ship and sea rescue service
   - Field hospital
2. Departmental managers:
   - Doctors
   - Veterinary doctors
   - Dentists
   - Planning
3. Executive Manager:
   - The doctor general

The IRS-process with selection of qualified and competent participants in interviews is coordinated in a cross-organizational DHS IRS Reference group. Resulting documentation and method support is delivered by an external facilitator and an experienced DHS project manager in an IRS Workgroup supported by the DHS IRS Reference group.

F2.3.2 Interviews

Each interview has the following participants:

- Interviewers (project manager and facilitator)
- A person from the IRS reference group
- Selected persons from section or management level

The interview participants have the responsibility for the quality of the report content.

We have established five sectional reports with the following content:

- Tasks, products, and objective
- Why the tasks are performed in the context of DHS
- Information used
- Information not available that could improve section performance
- Information exchange with others
- Possible improvements of current systems and procedures
- Improvements of current systems and procedures from integration of systems
- Potential benefits from improved systems, procedures, and integration
- Suggested solutions (not required)

We have established three departmental reports:

- Department responsibility and tasks
  - Who are the primary clients?
  - Who are the primary vendors?
- The results and the result quality (client needs satisfied) of the departments activity
- Expected and required quality of the departments products
- Need of improved information
- Potential improvement of decision foundation
- Expected benefits from improvements

We have established one executive level report (doctor general):

- The business objectives of DHS
- Need of improved information
- Potential improvement of the foundation for decision making
- Expected benefits from improvements

F2.3.3 Consolidation of the Interview Reports

The conclusions in the interview reports are coordinated in a workshop with the IRS Workgroup and the IRS Reference group that could have lost participants or have had participants added.

The interviews have shown a series of information elements used in the work of the interviewed:

- Some information elements describe actions (consultation, diagnosis, usage of material, experience element, etc.)
- Some information elements describe the DHS structure (organization, infirmary, material, patient, personnel, etc.)

In the Consolidation Workshop the participants decide on the functionality that is required in the future information system. The point of departure is the actions of DHS and these actions' requirements for procedures and information in order to be performed efficiently. It is thus decided what functionality the system elements must offer to the users thereof.

This report is the result of the IRS interviews and the IRS Consolidation Workshop.

F2.4 Other Source Material

The following supplementary source material has been used:

- Pertinent civil and military healthcare legislation and conditions
- Electronic healthcare record architecture
- Civil healthcare classification system
- The ministry of defense IT-strategy

F3. FUNCTIONAL AREA INFORMATION NEEDS

FOSIS IRS has been conducted within the following organization structure:

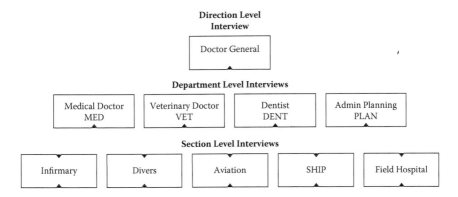

Sections refer to a military command unit, but for their rules of healthcare conduct they refer to DHS.

All managers have received all section reports before their IRS management level interview.

The doctor general had access to all sectional and departmental reports before the direction level interview.

F3.1 The Doctor General

F3.1.1 The Business Objectives of DHS

F3.1.2 Need of Improved Information

F3.1.3 Potential Improvement of the Foundation for Decision Making

F3.1.4 Expected Benefits from Improvements

By a more systematic follow-up on diagnosis and related treatments for the single patient, the doctor general would get a better foundation for inspections, perform better inspections, and get an improved foundation for continuous performance improvement.

Objective measurements of product quality can be used for improved quality management.

Education of healthcare personnel can be adapted to the epidemiological development pattern in the military, which will improve the performance of the education and the personnel in support of better infirmary performance.

Ongoing updated information about personnel education needs and wishes can improve the planning and performance of the healthcare personnel training and the healthcare personnel carrier opportunities.

If DHS gets the resource management responsibility for healthcare personnel, this management results in higher DHS performance.

The DHS product and performance can be made more visible by registration of time used in defense and civil research, which can contribute to a more realistic budget for DHS based on the real defense benefits from the DHS organization.

F3.2 Medical Doctors

The medical doctors have the responsibility for the defense healthcare quality.

F3.2.1 Products

The medical doctors establish qualitative and quantitative requirements to the healthcare condition in Army, Navy, and Air Force.

The structure of infirmary duty is adapted to the ongoing structural changes in the defense, where ordinary military personnel and personnel such as pilots and divers with special healthcare requirements are the primary clients (patients).

F3.2.2 Objective

The medical doctors must within their resource budget and the DHS conditions in general manage the medical treatment of the defense personnel in order to ensure the physical and psychological health of the personnel that best possible satisfies the defense requirements for functionally capable personnel.

F3.2.3 The Results and the Result Quality (Client Needs Satisfied) of the Departments Activity

The medical doctors give advice to the doctor general concerning:

- Defense healthcare needs
- Defense healthcare human resource requirements

The medical doctors coordinate the contact with civil national and international organizations.

Other responsibilities comprise:

- Personnel
- Inspection
- Infirmary capacity planning
- Education, skill, and competence planning

F3.2.4 Suggested Procedure Improvements and Their Benefits

- Improved skill and competence information for improved education planning and better education performance
- Improved information about military exercises placement and timing will make it possible to ensure a better service in these situations
- The patient health journal must always be accessible under the medical contact with the personnel, which will require an IT-supported patient journal system
- Only WHO-coding of diagnosis should be used for homogenous registration in the infirmaries and field hospitals
- FOSIS must be integrated with similar NATO-based systems for improved international cooperation under international crisis and war conditions with, for example, UN or NATO involvement
- Improved personnel information exchange between defense and civil doctors can provide much better healthcare to the concerned personnel

F3.3 Veterinary Doctors

F3.4 Dentists

F3.5 Planning

F3.6 Infirmaries

F3.7 Aviation medicine

F3.8 Diver Medicine

F3.9 Field Hospital (FHOSP)

An international FHOSP has been used as a model for this section. The relationships among departments inside the FHOSP and the relations with direct partners (the logistics battalion), other sanitary units (1.-3. og 5. echelon), other FHOSP, NATO-partners, DHS, and civil hospitals has been studied.

FHOSP is subordinated a logistics battalion. Replenishment of material and equipment and reparation and maintenance support is required through this logistics battalion.

FHOSP must be able to serve nationally and internationally, for example, as demanded from UN, NATO, OSCE, or EU partners.

FHOSP is focused on chirurgical treatment at least on high civil quality level. It receives, treats, and evacuates patients. It delivers sanitary equipment and material to sanitary units closer to the front (1.-3. echelon).

FHOSP comprises the following "departments":

- Reception and registration
- Ambulatory
- X-ray, EKG
- Dental clinic
- Operation rooms
- Intensive/anesthetic
- Bed sections
- Isolation
- Sterilization
- Laboratory
- Evacuation/leave instruction/administration/conference
- Sanitary supply
- Kitchen/cafeteria
- Operational (logistics and communication)
- Technique and maintenance (power, plumbing, electronics)
- Personnel quarters (tents, containers)
- Rolling materiel (parking)

FHOSP is 4. echelon, while the supply of material, power, plumbing, etc. is 1. echelon relative to FHOSP functionality. This means that basically all supply of material and equipment must be delivered from national ground because 1. echelon carries as little material as possible.

For the FHOSP a manual journal system has been developed. The field journal has been replaced by a Field Medical Card (FMC) covering the journaling needs at 2. and 3. echelon. FMC complies with NATO STANAG requirements. FMC is relatively fail-safe as no written documentation is required except for y/n tick off.

The route of the wounded is:

4. Place of injury (1. echelon)
5. Departmental place of bandage (2. echelon)
6. Main place of bandage (3. echelon)
7. FHOSP (4. echelon)
8. Civil hospital or garrison infirmary (5. echelon)

Healthcare documentation comprises:

1. Echelon—Fills out FMC in English, French, and national language.
2. Echelon—Fills out FMC front page in English, French, and national language.
3. Echelon—Fills out FMC last (back) page in English, French, and national language.
4. Echelon—From 4th echelon the language is national in journals and other documents. Documents to follow the patient through NATO links are in English.

FHOSP maintains a list of all patients who have been received for treatment.

The list from the last 24-hour period is sent to the operative command unit by military mail and is one of the foundations for requirement of replacement personnel, for the activation of sanitary units, and finally to be able to trace personnel.

F3.9.1 Products

FHOSP services always to be delivered at civil quality comprise:

- Chirurgical treatment (primarily)
- Observation and treatment
- Sanitary material supply
- Requisition of sanitary material
- Normal infirmary service to own personnel
- Administrative duties and communication

F3.9.2 Objective

FHOSP must treat sick and wounded with the intention to protect life and mobility. FHOSP must communicate to ensure optimal logistics of patients and of sanitary material and equipment. It is a national political requirement that FHOSP can be activated fast worldwide as part of a purely humanitarian effort.

F3.9.3 Work Processes and Procedures

The work processes and procedures for 1.-4. Echelon structure with FHOSP are:

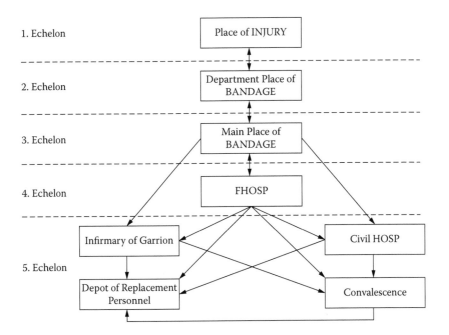

Place of Injury

FMC is filled in with:

- Person ID
- Grade
- Unit
- Name
- Time of injury
- Type of injury
- Contamination
- Intermediate diagnosis
- Transport Instruction

FMC follows the patient to 2. and 3. echelon.

Field journals such as the FMC are different from nation to nation.

Department and Main Place of Bandage

FMC is filled in with big differences in quality based on:

- Personal qualifications FMC
- National differences in format
- National differences in attitude

F3.9.3.1 Reception on FHOSP

F3.9.3.2 Treatment on FHOSP

F3.9.3.3 Evacuation from FHOSP

F3.9.3.4 Supply Logistics

F3.9.3.5 Patient Logistics

F3.9.4 Suggested Improvements to Procedures

- Access to centrally controlled and valid patient healthcare information
- Possibility to register validated patient healthcare information de-centrally with automatic central update when connected

F3.10 SHIP

F4. FOSIS SYSTEM REQUIREMENTS

By the implementation of healthcare information systems in the defense, two central tasks must be completed:

1. As IT and information systems are responsible, one must ensure that the future system can operate within the framework of the defense IT standard platforms. Even though you might be able to find and acquire suitable COTS products, these must be adapted to meet the DHS needs. This adaptation requires set up, development, and

system testing, integration testing, and acceptance testing of the COTS-based solution, which is done in a phase after the purchase of the COTS product.

2. As a user, you must ensure the ease of use of the future system, its ability to protect confidential patient information, its ability to ensure the ethical behavior of the personnel, and its ability to meet the defense security requirements. These properties cannot be controlled solely with IT. They require better and appropriate workflows, new personnel agreements, and training that is adapted to the improved behavior.

This consolidated IRS report not only describes the requirements to an IT COTS-based solution, but also describes the requirements of better DHS business workflows.

With FOSIS as it is outlined in this report, the defense will be able to contribute to an improvement of the public and private healthcare service.

F4.1 General FOSIS System Requirements

Confidential information.

F4.2 FOSIS Information System Modules

FOSIS must offer the following information system modules:

- Consultation
- Treatment
- Patient logistics
- Personnel planning
- Procurement
- Inventory management
- Development
- Maintenance
- Inspection

In the following chapters, the system module functionality is outlined.

For each module, selected parts of the overall information model will be presented to illustrate the objects used in the module. ACTION objects are shown with bold frame, while STRUCTURE objects are shown with normal frame.

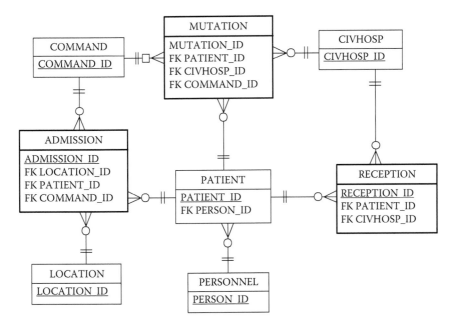

This module must ensure that the defense always knows where a patient is.

F4.3 Consultation

F4.4 Treatment

F4.4.1 Plan Treatment

F4.4.2 Perform Treatment

F4.4.3 Invoice to Civil Client or from Civil Treatment Vendor

F4.4.4 Pay Invoice

F4.4.5 Integration with Other Systems

F4.4.6 Target Users

F4.5 Patient Logistics

F4.5.1 Receive Patient

F4.5.2 Admit Patient

F4.5.3 Move Patient

F4.5.4 Exit Patient

F4.5.5 Integration with Other Systems

F4.5.6 Target Users

F4.6 Personnel Planning

F4.7 Procurement

F4.8 Inventory Management

F4.9 Development

F4.10 Maintenance

F4.11 Inspection

F5. CONCLUSION AND RECOMMENDATIONS

F5.1 Cost/Benefit Analysis

The cost and benefit analysis is based on rough cut estimated amounts because there are no statistical reports or accounts that can support an objective economic feasibility study of the introduction of FOSIS except for the few key figures shown below.

All amounts are determined on the basis of the benefits that the defense as a whole can achieve by improving the healthcare systems. Where these benefits are expressed as a public benefit of less strain on the public health-care system, these benefits are included. The immediate personal benefits that patients, other treated persons, and FSU personnel may obtain have not been included.

Some statistics have a relationship with FOSIS implementation benefits, for example:

- The current annual costs of DHS are close to €25 million.
- Soldiers discarded after acceptance to be employed by the defense cost €10,000 per person. This happens approximately 550 times per year with an annual cost of €5 million.

If one would try to estimate the real public benefits of FOSIS one could show astronomic amounts from:

- Reduced risk of cardiovascular diseases based on improved defense healthcare statistics
- The value of less F16 accidents annually
- The value of 50% less work injuries in the defense annually

The accumulated value of these benefits is so big that one can wonder why FOSIS has not been implemented a long time ago.

F5.1.1 Calculation of Benefits

Benefits	Value €/year
Maintenance of current systems	40,000
Improved prevention of personnel health problems (also useful for civil authorities)	150,000
...	150,000
...	40,000
...	...
...	...
Total €/year	5,000,000

F5.1.2 Estimation of Costs

Investment	€
Purchase of COTS for FOSIS implementation	800,000
Implementation of FOSIS workflows and reports	1,000,000
Implementation of FOSIS requirements in COTS	300,000

Investment	€
Development of training material	150,000
Design and development of integration with defense and civil systems	300,000
IT infrastructure and IT support organization development	400,000
User documentation	150,000
Initial training	150,000
Total	**3,250,000**

Annual costs	€
...	...
Total	**1,000,000**

F5.2 Recommended Information Systems and Their Priority

6. Consultation

   Treatment

7. Patient logistics
8. Personnel planning
9. Procurement

   Inventory management
   Material maintenance
   Inspection

10. Development

F5.3 Suggested Implementation Project

F5.3.1 Important Phases and Milestones

- Legally compliant tendering material for COTS-based FOSIS solution (turnkey)
- Detailed requirements spec for FOSIS as basis for turnkey agreement
- Legally compliant solicitation
- Selection of maximum five potential vendors
- Distribution of tendering material with turnkey contract
- Evaluation of offers and choice of vendor if possible
- Preparation of infrastructure for installation of basis COTS
- COTS installation and product sign off from defense IT
- PQA in the FOSIS IRS Reference group
- Selection of implementation project manager for the FOSIS solution

- Agile FOSIS development and implementation including documentation
- Implementation of FOSIS test and training environment
- Preparation of testers and test scenarios for Simulated Accept Test (SAT)
- FOSIS IRS Reference group accept test of FOSIS
- Establish learned lessons collection and communicate them

Attachment 1 Critical Success Factors
CSFs were formulated under PQA by the IRS Reference group:

1. Essential FOSIS functionality implemented simultaneously on time.
2. Intuitive Danish language user interface.
3. FOSIS supports all healthcare services throughout.
4. FOSIS provides access to necessary and complete healthcare information.
5. FOSIS communicates with relevant systems.
6. FOSIS is aligned with the defense IT strategy and is based on relevant standards.
7. FOSIS meets all requirements for safety and traceability.
8. FOSIS enables a flexible, user-specific data handling.
9. FOSIS increases quality and efficiency in healthcare service.

Attachment 2 Object Lifecycle Matrices (CRUD)
Note: C: Create, R: Read, U: Update, D: Delete.
Consultation:

Process Object	Plan Consultation	Perform Consultation	Invoice	Pay Invoice
DEPARTMENT	R	R		
INSTALLATION	R	R		
TREATMENT		R		
CIVIL INSTITUTION	R	R	R	R
DIAGNOSIS		C	R	
DIAGNOSIS CODE		R		
DOC REF	R	RC		
DOCUMENT	R	RC		
SUBJECT		R		
INVOICE			CU	CU
COMPONENT	C	RUC	R	R

Process Object	Plan Consultation	Perform Consultation	Invoice	Pay Invoice
STOCK LOCATION		R		
MATERIAL		R		
MATERIAL USE		C	R	
MYNDIGHED	R	R	R	R
CONSULTATION	C	RUC	R	R
...				
ORGANIZATION		R		

Treatment:

Patient logistics:

Personnel planning:

Procurement:

Inventory management:

Development:

Inspection:

Attachment 3 Object Descriptions

1. Consultation

Object Descriptions A = ACTION S = STRUCTURE			Used in Module									Form Page
* = used in module	A	S	4.3	4.4	4.5	4.6	4.7	4.8	4.9	4.10	4.11	Ref
DEPARTMENT		S	*	*	*		*	*			*	
INSTALLATION		S	*	*	*		*	*		*		
TREATMENT	A			*	*	*		*				
TREATMENT SCHEDULER	A			*								
PROJECT		S	*						*			
PROJECT GUIDE		S	*						*			
CLAIM	A					*						
...												

   2. User guide to the object description usage
   3. Object description columns user guide
   4. Object descriptions for FOSIS

Attachment 4 Input-Output Tables

Here is shown the total interchange of information between sections and sections and between sections and external organizations as this was documented in the sectional reports; including the information maintained in each section.

Table over information to and from IBA sections and externals

TO FROM	INF	FHOSP	SHIP	ORG
INF	Exit letter		Healthcare journal extract	
FHOSP				
SHIP			Approval of healthcare journal	
...				

Attachment 5 Vocabulary

Acronym Concept	Definition	Example
LOGBTN	Logistics battalion	
COMEDS	Committee of the Chiefs of Military Medical Services in NATO	
COTS	Commercial Off The Shelf	
EDI	Electronic Document Interchange	
FHOSP	Field Hospital	
WHO	World Health Organization	
WONCA	The WONCA International Classification Committee (WICC) has produced the International Classification of Primary Care (ICPC), a clinical coding system for primary healthcare.	World Organization of National Colleges, Academies (WONCA) of doctors
...	...	...

Attachment 6 Key Figures

1. Infirmaries

INF activity:

	1993	1994	1995	1996
Medical				
Consultations	205.266	173.636	200.391	
Treatments	34.074	32.636	38.772	
Dental clinic				
Consultations	142.251	153.331	140.425	158.359
	1993	**1994**	**1995**	**1996**

Treatments	55.209	52.720	45.928	47.523
Working days			12.152	11.256
External consultations	53.417	53.486	40.191	39.322
No shows	7.717	7.749	6.120	5.684

2. Diver medicine
3. FHOSP
4. SHIP

Attachment 7 ERD User Guide

Appendix G: Order Handling IRS Section Report Example

G1. PARTICIPANTS (EXPERIENCE, INTERVIEWERS)

G1.1 Experience

Frantz M, RM (Relationship Manager)
Olga P, RM
Carl M, PM
Tom H, ED (Execution Desk)
Sylvie P, MO (Middle Office)

G1.2 Interviewers

IRS Reference
PM
Facilitator
Key Figures 20 Relationship Managers distributed by geographical region
200 new clients per year
60 clients closing their accounts
Client call frequency (3–5% daily, 10–20% weekly, 50% monthly, 100% 6 months)
15% (ca. 400) managed portfolios (relatively big)
85% RM controlled portfolios
20 standard portfolios correspond to Benchmark
Bloomberg Security Prices 1 day old

G2. SECTION SCOPE, PRODUCT, AND PURPOSE

Current interview was conducted in the IRS Section comprising all departments/functions occupied with the execution of an order, for example:

- PM (Portfolio Manager) performing enhancements to managed portfolios
- PM monthly preparing list of recommended products
- RM contacting or being contacted by a client in order to improve the client's portfolio
- MO preparing the execution of orders or actually executing fund orders and small FOREX
- Small FOREX is less than $20,000 USD
- MO handling cash transactions in and out of portfolios
- ED making agreements with brokers
- ED executing all orders and FOREX

G2.1 Main Functions

Department Function	PM	RM	MO	ED
Prepare Client Profile		*		
Evaluate Securities Markets	*			
Prepare Recommended Products	*			
Prepare and Evaluate Portfolio Strategy	*	*		
Evaluate Portfolio Performance	*	*		
Lombard			*	
Transfer Cash in/out of Portfolios			*	
FOREX			*	*
Cash Deposits			*	*
Prepare Security Order		*	*	
Execute Security Order				*
Execute Fond Order			*	
Execute Security Transfer (Close Account)			*	
Deliver Ad Hoc Portfolio Reports to Client			*	

G2.2 Organizational Relationships

PM, RM, or MO prepares orders with anonymous client information and security information for handling by ED.

ED communicates with brokers and banks in order to purchase or sell securities, funds, currency positions, and options on the best possible terms.

MO communicates with CSAM concerning fund deals.

MO uses Bloomberg exchange rates for small FOREX deals.

ED handles all other FOREX and negotiates more favorable rates directly with banks.

PM uses information from HQ, Bloomberg, and many other sources for the preparation of the monthly Recommended Products List.

PM uses client information from RM in order to establish the best possible portfolio strategy for the client.

RM communicates with the client in order to get a complete understanding of all the decision parameters used by the client. The client attitude is highly dependent on the market situation, the performance of the client's profile, and the client's own situation.

G2.3 Product and Purpose

RM establishes a client profile without revealing the identity of the client. Only the RM knows all decision parameters of a client and only a limited set of these parameters are visible in the client master data. The most complete knowledge about the client's profile (age, citizenship, occupation, country of origin, requirements regarding access to the portfolio values, current risk profile, etc.) is required in order to be able to establish the best possible portfolio profile for the client.

A client's portfolio can contain any combination of equities, equity equivalents (options), bonds, bond equivalents, cash, currency positions, and precious metals (silver, gold) eventually partly financed with loans.

RM and PM maintain the portfolio combination of securities, etc. in such a way that it complies with the requirements of the client and with the overall risk policy.

Currently the RM and PM recommend and choose primarily securities with bank coverage, which gives the highest probability to fulfill the client's expectations.

In the longer term, the bank will establish strategies, which target on having a certain percentage of given products in all portfolios. Such products can be Protected Investment Notes (10-year certificate with long/short strategy with capital protection) or Alternative Performance Certificates with long/short strategy without capital protection, for example, selected based on the age of the client.

The general attitude is that what is good for the client is also good for the bank. The bank would never expose a client to unnecessary risk or recommend less optimal behavior by the client in order to optimize short-term profits. Currently this policy is visual in the fact that the cost of a deal to the bank and to the client in terms of commissions and/or curettage is not visible to RM, MO, or ED during order processing. ED can negotiate special commission conditions for major deals, but normally negotiated conditions apply. RM can give clients special conditions, but again normally standard conditions apply.

The overall purpose is clearly to optimize the portfolio performance within the constraints defined by the RM and the client.

RM receives commission based on both requisition and revenue objectives.

G3. AVAILABLE INFORMATION AND ITS USAGE

G3.1 Prepare Client Profile

RM is the only link to the client.

RM develops client information in communication with the client. The RM acquires and collects the material, which identifies and describes the client, and which defines the client margins (commission on transactions and interest difference from central bank rate on current accounts). A client can have very specific commission conditions, which are marked with a code on the client checklist.

CSU uses the paper-based client checklist information from RM for input into client files, which are protected from access.

RM maintains personal client information in a paper-based client file. LCD approves the client information and the client account opening.

Client information relating to portfolio handling (deals, corporate actions) is the agreement on margins (tariffs) applied to the different transactions. This information is used in the system to calculate and book commission for contribution calculation in accounting.

RM e-mails changes to client static data to MO.

RM can make a pledge agreement with a client, which allows PB to sell securities in order to ensure that the client's pledge value is not exceeded in situations with declining portfolio position values (client credibility).

G3.2 Evaluate Securities Markets

PM surveys the securities markets using analysis and recommendations from HQ and other available sources. Security price information is updated by BO (Back Office) Static Data automatically from Telekurs (BO corrects

Telekurs transfer errors manually). These errors are caused by securities traded on more than one stock exchange in more than one currency.

G3.3 Prepare Recommended Products

PM produces a list of recommended securities monthly to the RM.

G3.4 Prepare and Evaluate Portfolio Strategy

PM maintains standard portfolio profiles corresponding to different portfolio strategies relating to the client risk profile. There is a portfolio strategy for each supported currency ($, €, SFR) and some strategies have mixed currencies.

PM maintains managed portfolios risk profile and portfolio strategy.

RM handles the client risk profile concerning the portfolios controlled by the client. PM can support the initial establishment of a portfolio.

G3.5 Evaluate Portfolio Performance

PM ongoing evaluates that their managed portfolios perform according to the established strategy by controlling that the combination of portfolio positions comply with the strategy.

In cases of portfolio non-compliance with the strategy, positions are changed by selling and purchasing positions.

G3.6 Transfer Cash In/Out of Portfolios

RM controls all cash delivered by a client to his portfolio. The client must prove that the cash is legally obtained according to the control rules concerning white washing. It is practically impossible to get cash approved according to these rules. Normally only registered bank checks or transfers from a client-owned bank account are accepted. LCD must approve in-payments directly from clients.

Cash is normally received by SWIFT in BO, who prints the SWIFT message and sends it to MO for booking.

MO books the accepted cash on the client portfolio against NOSTRO.

MO validates outgoing payments, which are booked by Cashiers Desk.

Cash is normally sent by SWIFT in BO.

BO handles payments concerned with settlement.

RM (or PM concerning managed portfolios) handles client's requirements for cash payments (transfers). BO (SWIFT) does the physical transfer or, in rare cases, the Cashiers Desk does it.

G3.7 FOREX

ED performs major FOREX deals (>200,000) and will often negotiate special exchange rate with counterpart (most often a bank).

MO performs smaller FOREX deals using Bloomberg standard exchange rates.

SEM is used for FOREX.

G3.8 Cash Deposits

RM, PM, and MO initiate cash deposits (on behalf of clients):

- Manual deposits are not renewed automatically.
- Automatic deposits are moved to a new period automatically.

MO books the deposit on the client portfolio. MO sends a list of manual deposits to the RM every day.

ED makes the deposits per currency defining rate, period, currency, and margin every day.

G3.9 Prepare Security Order

RM prepares all types of security orders except for funds. All RM communication with the client is recorded on tape.

MO prepares all types of security orders received in writing from a client and fund orders.

PM prepares all types of orders concerning managed portfolios.

Purchase and Sale Security Order types comprise:

- Equity and equity equivalents
- Bonds and bond equivalents
- Derivatives

The static security data in SD is the primary source for security information, but the information system does not comprise complete information. Telekurs transmitted list with end of day prices from the day before are available to RM, MO, and PM.

SD only contains limited client information and no portfolio information. No one has access to complete client information because much client information is hidden in personal paper-based files.

RM and PM use primarily Bloomberg updated prices (these are 1 day old) or they get spot prices directly from the used broker or depositary. RM and PM do not have online access to real time security prices, but they do have access to Bloomberg dealable prices, which brokers are willing to deal at.

RM, PM, and MO book their security order in SD, which generates a window with the order at ED. Basically the transaction and price is not 100% safe before the order has been validated, reconciled, and approved in BO prior to final settlement, which is also done by BO.

RM, MO, PM, or ED cannot see the full contribution from an order before it has been booked and approved in BO.

G3.8.1 New Securities in SD

In cases where the security is not in SD, the order is executed with a preliminary (provisional) value number (static security data have two alternative keys: internal number and ISIN code, where the ISIN code is the link to external information from, e.g., Bloomberg and Telekurs).

In all circumstances, the order is written out on paper by ED (or MO for fund orders) and e-mailed to BO, where it is validated, where detailed information is added in IBSY, and where it is finally approved and settled.

G3.8.2 Derivatives

Derivatives such as OTC are used primarily to protect the value of portfolio positions and to increase the income from these positions. Commission on derivatives is much lower than on other security transactions.

Buying and selling derivatives follow the same workflow as other securities. The deviation from normal shares is that derivatives represent those and that the derivatives value depends on the development of the share price. Furthermore, the execution conditions are different for European (can only be executed on the due date) and American (can be executed on any date before or on due date). Derivatives come in 2 types:

- Exchange options (stock prices available)
- Over the counter (OTC) issued through ING

BO handles execution of derivatives automatically. This is not visible in the portfolio.

Under normal circumstances derivatives are not executed. It is especially seldom that clients choose to execute a derivative, but it can happen under conditions with fast rising or falling share prices. In order to be more reactive to execution opportunities and threats, a new list showing development in underlying share prices is currently under internal development. RM informs the client about execution of client portfolio derivatives based on information from BO. If a client asks RM to execute a derivative, BO performs the actual procedure after RM request.

Derivative Executions are time-consuming and there are examples where underlying shares have been traded when they have been used for execution.

G3.8.3 Limited Orders

Limited orders are time-constrained and price-constrained sales or purchase orders. Limited orders are automatically deleted after deadline.

G3.9 Execute Security Order

The ED who expedites the order from RM, MO, or PM tells this to the other ED by yelling (in order to avoid that other ED starts expediting the same order).

MO executes orders from HQ and funds.

ED reports the agreed security price and other conditions back to the RM by e-mail or phone (normally within a few minutes).

The RM can then confirm the price to the client. The RM checks the cash position of the client in IBSY. Sometimes actually received cash has not been booked in time and is therefore not visible to the RM. Lombard is not visible on the client portfolio.

G3.10 Execute Fund Order

Fund orders are prepared and executed by MO.

G3.11 Execute Security Transfer (Close Account)

MO handles the preparation of security transfers out of portfolios when accounts are closed or caused by, for example, corporate actions or simple collection of same security on same depository on a portfolio.

When account closings are involved, MO must have access to the original instruction.

The transfer is documented on paper, which is sent to BO for the physical handling of the transfer (SWIFT). Afterward, MO verifies that BO acted as required.

G3.12 Deliver Ad Hoc Portfolio Reports to Client

MO produces ad hoc portfolio reports to clients.

G4 REQUIRED INFORMATION, CURRENTLY NOT AVAILABLE

G4.1 Prepare Client Profile

Client conditions (margins) are showed only by codes, which are not easily memorized. Code translation is only available on Excel spreadsheet.

Complete client relationship documentation is not available in an easy accessible form for RM although all client RM communication is recorded.

Documentation used or produced in connection with client relationship incidents are not logged with reference to this incident.

G4.2 Evaluate Securities Markets

G4.3 Prepare Recommended Products

G4.4 Prepare Portfolio Strategy

G4.5 Evaluate Portfolio Strategy

G4.6 Evaluate Portfolio Performance

Only limited time series data is available for performance statistic calculation (e.g., it is impossible to separate the different movement types [value gains, capital change, dividends/interests] on portfolios without using artificial pricing in connection with free transfers and corporate actions such as dividends paid with shares).

G4.7 Transfer Cash In/Out of Portfolios

Cash is normally received by SWIFT in BO, who prints the SWIFT message and sends it to MO for booking. This procedure is time-consuming and can result in wrong cash availability on a portfolio, which can prevent an agreed deal from being approved.

G4.8 FOREX

G4.9 Cash Deposits

G4.10 Prepare Security Order

The RM or ED cannot see the calculation of the contribution from a given deal before the deal has been booked and approved (in BO).

The RM has no access to Lombard from the client portfolio, which means that the RM cannot control the client cash position completely.

RM does not get information about deletions of limited orders. RM should be triggered in order to be able to ask the client whether a prolongation or renewal is required.

G4.11 Execute Security Order

The Stock Exchange information in securities static data is not reliable, which can result in selection of a wrong depositary, which is very costly to correct.

A security should have only one depository account for the same portfolio.

G4.12 Execute Fund Order

G4.13 Execute Security Transfer (Close Account)

MO has no access to known planned corporate actions while closing an account and transferring its positions.

G4.14 Deliver Ad Hoc Portfolio Reports to Client

The delivered portfolio evaluation from MO is often not correct and does not correspond to what the RM can see.

G5. VALUE OF REQUIRED INFORMATION

The RM access to timely and reliable information about security prices and client cash position gives the RM much better opportunities to react fast to client transactions requirements.

Timely information about derivatives execution risk gives a better understanding of the client portfolio risk situation and makes it possible to react fast on execution risks and opportunities.

Consistent, complete, and valid information on portfolios and about securities makes it possible to react faster and more precise on the market development in order to optimize the portfolio performance.

G6. INFORMATION FLOW

To Order Handling (departments)	From Department/External
Client Order (RM, MO)	Client
Account Closing	RM, LCD
Order Approval	BO
Security	BO
CASH TRANSACTION	Client
CASH DEPOSIT (RM, MO)	Client
loan request (RM)	Client
Client (RM)	Client
Account Opening	LCD
Order Confirmation (ED)	Broker
Maintained by Order Handling (departments)	
Client (RM)	
Standard Portfolio (PM)	
CASH DEPOSIT (ED)	

From Order Handling (departments)	To Department/External
Client Order (RM, MO)	ED
Client Order (ED)	BO
ORDER CONFIRMATION	Client
CASH DEPOSIT (RM)	Client
MANUAL CASH DEPOSIT (MO)	RM
Loan Request (RM)	MO
Client (RM)	CSU, LCD
Account Opening Request (RM)	LCD
Order Confirmation (ED)	BO, RM

G7. SUGGESTED ENHANCEMENTS TO (IT) SYSTEMS

Client information should be complete on IT especially for RM, while protected from other access.

All client communication should be logged and available to RM, as well as communication between client and PM or BO, when this is relevant.

The minimum fee on security transactions is not automatically guaranteed.

The e-mail information to RM is not smart. RM needs complete cash position and portfolio evaluation for a client at any time.

The reconciliation of cash transactions (especially received checks) takes too long, which gives unreliable cash position information from the portfolio. RM has no access to transaction details, but MO does.

MO can see a corporate action, but the corporate action is not related to the underlying security, which should be the case.

Derivatives must be handled correctly in IT in order to avoid deals on underlying positions in execution situations.

USWHT must be handled correctly in IT avoiding incorrect bookings and tax deductions in connection with corporate actions and deals.

G8. SUGGESTED INTEGRATION OF SYSTEMS

RM needs faster access to realistic or guaranteed security prices in order to be able to respond immediately to client demands.

DS should be integrated with IT for full cash position on portfolios.

SD and IT should be better integrated with well-defined common data avoiding unnecessary double maintenance of the same information and errors.

Corporate actions must be visible on portfolios as soon as they are reported to BO and registered in order to avoid erroneous transactions on involved securities.

G9. EXPECTED BENEFITS FROM ENHANCEMENTS

Better portfolio performance (PM, RM, MO).

More satisfied clients based on faster access to reliable portfolio information such as full visibility of transaction status and planned corporate actions (RM, MO).

Better contribution from overall business activity based on better foundation for portfolio maintenance (PM, RM).

G10. SUGGESTIONS FOR (NEW) IT SOLUTIONS

New corporate action solution integrated with IT.

DS solution integrated with IT.

Index